OPTIMIZATION

PURE AND APPLIED MATHEMATICS

A Program of Monographs, Textbooks, and Lecture Notes

Contributions to *Lecture Notes in Pure and Applied Mathematics* are reproduced by direct photography of the author's typewritten manuscript. Potential authors are advised to submit preliminary manuscripts for review purposes. After acceptance, the author is responsible for preparing the final manuscript in camera-ready form, suitable for direct reproduction. Marcel Dekker, Inc. will furnish instructions to authors and special typing paper. Sample pages are reviewed and returned with our suggestions to assure quality control and the most attractive rendering of your manuscript. The publisher will also be happy to supervise and assist in all stages of the preparation of your camera-ready manuscript.

LECTURE NOTES
IN PURE AND APPLIED MATHEMATICS

1. *N. Jacobson,* Exceptional Lie Algebras
2. *L.-Å. Lindahl and F. Poulsen,* Thin Sets in Harmonic Analysis
3. *I. Satake,* Classification Theory of Semi-Simple Algebraic Groups
4. *F. Hirzebruch, W. D. Newmann, and S. S. Koh,* Differentiable Manifolds and Quadratic Forms (out of print)
5. *I. Chavel,* Riemannian Symmetric Spaces of Rank One (out of print)
6. *R. B. Burckel,* Characterization of C(X) Among Its Subalgebras
7. *B. R. McDonald, A. R. Magid, and K. C. Smith,* Ring Theory: Proceedings of the Oklahoma Conference
8. *Y.-T. Siu,* Techniques of Extension of Analytic Objects
9. *S. R. Caradus, W. E. Pfaffenberger, and B. Yood,* Calkin Algebras and Algebras of Operators on Banach Spaces
10. *E. O. Roxin, P.-T. Liu, and R. L. Sternberg,* Differential Games and Control Theory
11. *M. Orzech and C. Small,* The Brauer Group of Commutative Rings
12. *S. Thomeier,* Topology and Its Applications
13. *J. M. Lopez and K. A. Ross,* Sidon Sets
14. *W. W. Comfort and S. Negrepontis,* Continuous Pseudometrics
15. *K. McKennon and J. M. Robertson,* Locally Convex Spaces
16. *M. Carmeli and S. Malin,* Representations of the Rotation and Lorentz Groups: An Introduction
17. *G. B. Seligman,* Rational Methods in Lie Algebras
18. *D. G. de Figueiredo,* Functional Analysis: Proceedings of the Brazilian Mathematical Society Symposium
19. *L. Cesari, R. Kannan, and J. D. Schuur,* Nonlinear Functional Analysis and Differential Equations: Proceedings of the Michigan State University Conference
20. *J. J. Schäffer,* Geometry of Spheres in Normed Spaces
21. *K. Yano and M. Kon,* Anti-Invariant Submanifolds
22. *W. V. Vasconcelos,* The Rings of Dimension Two
23. *R. E. Chandler,* Hausdorff Compactifications
24. *S. P. Franklin and B. V. S. Thomas,* Topology: Proceedings of the Memphis State University Conference
25. *S. K. Jain,* Ring Theory: Proceedings of the Ohio University Conference
26. *B. R. McDonald and R. A. Morris,* Ring Theory II: Proceedings of the Second Oklahoma Conference
27. *R. B. Mura and A. Rhemtulla,* Orderable Groups
28. *J. R. Graef,* Stability of Dynamical Systems: Theory and Applications
29. *H.-C. Wang,* Homogeneous Banach Algebras
30. *E. O. Roxin, P.-T. Liu, and R. L. Sternberg,* Differential Games and Control Theory II
31. *R. D. Porter,* Introduction to Fibre Bundles
32. *M. Altman,* Contractors and Contractor Directions Theory and Applications
33. *J. S. Golan,* Decomposition and Dimension in Module Categories
34. *G. Fairweather,* Finite Element Galerkin Methods for Differential Equations
35. *J. D. Sally,* Numbers of Generators of Ideals in Local Rings
36. *S. S. Miller,* Complex Analysis: Proceedings of the S.U.N.Y. Brockport Conference
37. *R. Gordon,* Representation Theory of Algebras: Proceedings of the Philadelphia Conference
38. *M. Goto and F. D. Grosshans,* Semisimple Lie Algebras
39. *A. I. Arruda, N. C. A. da Costa, and R. Chuaqui,* Mathematical Logic: Proceedings of the First Brazilian Conference

40. *F. Van Oystaeyen,* Ring Theory: Proceedings of the 1977 Antwerp Conference
41. *F. Van Oystaeyen and A. Verschoren,* Reflectors and Localization: Application to Sheaf Theory
42. *M. Satyanarayana,* Positively Ordered Semigroups
43. *D. L. Russell,* Mathematics of Finite-Dimensional Control Systems
44. *P.-T. Liu and E. Roxin,* Differential Games and Control Theory III: Proceedings of the Third Kingston Conference, Part A
45. *A. Geramita and J. Seberry,* Orthogonal Designs: Quadratic Forms and Hadamard Matrices
46. *J. Cigler, V. Losert, and P. Michor,* Banach Modules and Functors on Categories of Banach Spaces
47. *P.-T. Liu and J. G. Sutinen,* Control Theory in Mathematical Economics: Proceedings of the Third Kingston Conference, Part B
48. *C. Byrnes,* Partial Differential Equations and Geometry
49. *G. Klambauer,* Problems and Propositions in Analysis
50. *J. Knopfmacher,* Analytic Arithmetic of Algebraic Function Fields
51. *F. Van Oystaeyen,* Ring Theory: Proceedings of the 1978 Antwerp Conference
52. *B. Kedem,* Binary Time Series
53. *J. Barros-Neto and R. A. Artino,* Hypoelliptic Boundary-Value Problems
54. *R. L. Sternberg, A. J. Kalinowski, and J. S. Papadakis,* Nonlinear Partial Differential Equations in Engineering and Applied Science
55. *B. R. McDonald,* Ring Theory and Algebra III: Proceedings of the Third Oklahoma Conference
56. *J. S. Golan,* Structure Sheaves over a Noncommutative Ring
57. *T. V. Narayana, J. G. Williams, and R. M. Mathsen,* Combinatorics, Representation Theory and Statistical Methods in Groups: YOUNG DAY Proceedings
58. *T. A. Burton,* Modeling and Differential Equations in Biology
59. *K. H. Kim and F. W. Roush,* Introduction to Mathematical Consensus Theory
60. *J. Banas and K. Goebel,* Measures of Noncompactness in Banach Spaces
61. *O. A. Nielsen,* Direct Integral Theory
62. *J. E. Smith, G. O. Kenny, and R. N. Ball,* Ordered Groups: Proceedings of the Boise State Conference
63. *J. Cronin,* Mathematics of Cell Electrophysiology
64. *J. W. Brewer,* Power Series Over Commutative Rings
65. *P. K. Kamthan and M. Gupta,* Sequence Spaces and Series
66. *T. G. McLaughlin,* Regressive Sets and the Theory of Isols
67. *T. L. Herdman, S. M. Rankin, III, and H. W. Stech,* Integral and Functional Differential Equations
68. *R. Draper,* Commutative Algebra: Analytic Methods
69. *W. G. McKay and J. Patera,* Tables of Dimensions, Indices, and Branching Rules for Representations of Simple Lie Algebras
70. *R. L. Devaney and Z. H. Nitecki,* Classical Mechanics and Dynamical Systems
71. *J. Van Geel,* Places and Valuations in Noncommutative Ring Theory
72. *C. Faith,* Injective Modules and Injective Quotient Rings
73. *A. Fiacco,* Mathematical Programming with Data Perturbations I
74. *P. Schultz, C. Praeger, and R. Sullivan,* Algebraic Structures and Applications: Proceedings of the First Western Australian Conference on Algebra
75. *L. Bican, T. Kepka, and P. Nemec,* Rings, Modules, and Preradicals
76. *D. C. Kay and M. Breen,* Convexity and Related Combinatorial Geometry: Proceedings of the Second University of Oklahoma Conference
77. *P. Fletcher and W. F. Lindgren,* Quasi-Uniform Spaces
78. *C.-C. Yang,* Factorization Theory of Meromorphic Functions
79. *O. Taussky,* Ternary Quadratic Forms and Norms
80. *S. P. Singh and J. H. Burry,* Nonlinear Analysis and Applications
81. *K. B. Hannsgen, T. L. Herdman, H. W. Stech, and R. L. Wheeler,* Volterra and Functional Differential Equations

82. *N. L. Johnson, M. J. Kallaher, and C. T. Long,* Finite Geometries: Proceedings of a Conference in Honor of T. G. Ostrom
83. *G. I. Zapata,* Functional Analysis, Holomorphy, and Approximation Theory
84. *S. Greco and G. Valla,* Commutative Algebra: Proceedings of the Trento Conference
85. *A. V. Fiacco,* Mathematical Programming with Data Perturbations II
86. *J.-B. Hiriart-Urruty, W. Oettli, and J. Stoer,* Optimization: Theory and Algorithms

Other Volumes in Preparation

OPTIMIZATION

Theory and Algorithms

edited by

Jean-Baptiste Hiriart-Urruty
Laboratoire d'Analyse Numérique
Université Paul Sabatier
Toulouse, France

Werner Oettli
Universität Mannheim
Mannheim, West Germany

Josef Stoer
Universität Würzburg
Wurzburg, West Germany

MARCEL DEKKER, INC. New York and Basel

Library of Congress Cataloging in Publication Data
Main entry under title:

Optimization : theory and algorithms.

(Lectures notes in pure and applied mathematics ; v. 86)
1. Mathematical optimization--Addresses, essays, lectures. I. Hiriart-Urruty, Jean-Baptiste
II. Oettli, Werner. III. Stoer, Josef.
IV. Series.
QA402.50674 1983 511 83-7599
ISBN 0-8247-7019-6

MARCEL DEKKER, INC.
270 Madison Avenue, New York, New York 10016

Current printing (last digit):
10 9 8 7 6 5 4 3 2 1

PRINTED IN THE UNITED STATES OF AMERICA

PREFACE

From March 16 to 20, 1981, a conference entitled "Optimization: Theory and Algorithms" was held in Confolant (Puy-de-Dôme, France). This meeting, which was directed by the editors of this volume, was devoted to the theory of Optimization in the broad sense of the word. It was attended by 72 participants mainly from France and West Germany but also from 9 further countries. Its format was thirty five 40-minute lectures by speakers who developed aspects of their best current research. All the participants concurred in the usefulness of such a gathering which brought scolars together from different areas of Optimization.

This volume contains a collection of 15 papers presented at the conference. It is divided into three parts. The first part contains 7 papers which concern mainly the *theory*; it includes papers on tangent cones, duality formulas, a generalized concept of conjugation, the notion of maximinimizing sequence for a saddle-point problem, the continuity properties of marginal functions in parametric programming, a new version of the Hahn-Banach principle, and the ε-approximation of a function in the Chebyshev norm. The second part deals more with *algorithms* in Optimization. It contains a first paper on the multiple exchange algorithm in convex programming, a second one on algorithms to extract convergent subsequences from a non-convergent sequence and a third one devoted to the n-step square convergence of some minimization procedures related to Powell's derivative free method. In the third part are shown various *applications* of results and techniques from Optimization to related areas. The 5 papers in this part include works on the optimal reconstruction of surfaces using parametric spline functions, the penalty methods for constrained variational inequalities in continuum mechanics, necessary optimality conditions and bang-bang principle for time-optimal parabolic boundary control problems with integral state constraints, optimality conditions for weight minimization of structures under dynamic load and problems of estimation and control in Markov decision models with finite state space.

We would like to express our gratitude to The Scientific Council of the University of Clermont-Ferrand II, The French Mathematical Society (S.M.F.) which sponsored the conference and provided a financial support, The European Research Office of the United States Army (USARDSG-UK) and the European Office of Aerospace Research and Development of the United States Air Force (EOARD) which generously provided a financial support under grant DAJA37-81-M-0188 *.

Finally, we would like to thank the referees and the typists for their invaluable help during the preparation of this volume.

Jean-Baptiste Hiriart-Urruty
Werner Oettli
Josef Stoer

* The views, opinions, and/or findings contained in this report are those of the authors and should not be construed as an official Department of the U.S. Army/Department of the U.S. Air Force position, policy, or decision, unless as designated by other documentation.

CONTENTS

PREFACE iii

CONTRIBUTORS vii

CONFERENCE PARTICIPANTS ix

I. THEORY

HYPERTANGENT CONES FOR A SPECIAL CLASS OF SETS
S. Dolecki 3

OPTIMIZATION BY LEVEL SET METHODS I : DUALITY FORMULAE
I. Singer 13

A GENERALIZED CONCEPT OF CONJUGATION
J.-E. Martinez-Legaz 45

WELL-POSED SADDLE POINT PROBLEMS
E. Cavazzuti and J. Morgan 61

CONTINUITY PROPERTIES OF PERFORMANCE FUNCTIONS
J.-P. Penot 77

ON A GENERAL FORMULATION OF THE HAHN-BANACH PRINCIPLE WITH APPLICATION TO OPTIMIZATION THEORY
W. Oettli 91

A NOTE ON THE CHEBYSHEV ε-APPROXIMATION PROBLEM
J.-J. Strodiot, V. H. Nguyen, and N. Heukemes 103

II. ALGORITHMS

A MULTIPLE EXCHANGE ALGORITHM IN CONVEX PROGRAMMING
H.-P. Blatt, U. Kaiser, and B. Ruffer-Beedgen 113

ALGORITHMES POUR EXTRAIRE UNE SOUS-SUITE CONVERGENTE D'UNE SUITE NON CONVERGENTE
J.-P. Delahaye 131

THE n-STEP SQUARE CONVERGENCE OF SOME MINIMIZATION ALGORITHMS RELATED TO POWELL'S DERIVATIVE FREE METHOD
J. Stoer 153

III. APPLICATIONS

OPTIMAL RECONSTRUCTION OF SURFACES USING PARAMETRIC SPLINE FUNCTIONS
P. Cinquin 187

ON THE PENALTY METHOD FOR CONSTRAINED VARIATIONAL INEQUALITIES
J. Gwinner 197

BANG-BANG-CONTROLS FOR TIME-OPTIMAL PARABOLIC BOUNDARY CONTROL PROBLEMS WITH INTEGRAL STATE CONSTRAINTS
U. Mackenroth 213

STATIC AND DYNAMIC LOADS, POINTWISE CONSTRAINT IN STRUCTURAL OPTIMIZATION
B. Rousselet 225

ESTIMATION AND CONTROL IN FINITE STATE DISCOUNTED DYNAMIC PROGRAMMING
M. Schäl 239

CONTRIBUTORS

H.-P. BLATT,* Fakultät für Mathematik und Informatik, Universität Mannheim, Mannheim, West Germany

ENNIO CAVAZZUTI, Departimento di Matematica, Università di Modena, Modena, Italy

PHILIPPE CINQUIN, Département d'Informatique Medicale, Center Hospitalier Regional, Dijon, France

JEAN-PAUL DELAHAYE, Département d'Informatique de l'I.U.T., Université des Science et Techniques de Lille, Villeneuve d'Ascq, France

SZYMON DOLECKI,† Institute of Mathematics, Polish Academy of Sciences, Warszawa, Poland

J. GWINNER, Research and Development, Daimler-Benz AG, Stuttgart, West Germany

NORBERT HEUKEMES, Departement de Mathématiques, Facultés Universitaires de Namur, Namur, Belgium

---

* Current Affiliation - Katholische Universität Eichstätt, Eichstätt, West Germany

† Current Affiliation - Dipartimento di Matematica, Università di Trento, Trento, Italy

ULRICH KAISER, Fakultät für Mathematik und Informatik, Universität Mannheim, Mannheim, West Germany

V. MACKENROTH, Mathematisches Institut, Universität Bayreuth, Bayreuth, West Germany

JUAN-ENRIQUE MARTINEZ-LEGAZ, Departamento de Ecvaciones Funcionales, Facultad de Matemáticas, Universidad de Barcelona, Barcelona, Spain

JACQUELINE MORGAN, Istituto di Matematica, Facoltà di Scienze, Università di Napoli, Napoli, Italy

V. HIEN NGUYEN, Department of Mathematics, Facultés Universitaires de Namur, Namur, Belgium

WERNER OETTLI, Fakultät für Mathematik und Informatik, Universität Mannheim, Mannheim, West Germany

JEAN-PAUL PENOT, Département de Mathématiques et d'Informatique, Faculté des Sciences, Université de Pau, Pau, France

BERNARD ROUSSELET, Departement de Mathématiques, Université de Nice, Nice, France

B. RUFFER-BEEDGEN, Fakultät für Mathematik und Informatik, Universität Mannheim, Mannheim, West Germany

MANFRED SCHÄL, Institut für Angewandte Mathematik, Universität Bonn, Bonn, West Germany

IVAN SINGER, Department of Mathematics, INCREST National Institute for Scientific and Technical Creation, Bucuresti, Romania

JOSEF STOER, Institut für Angewandte Mathematik und Statistik, Universität Würzburg, Würzburg, West Germany

J.-J. STRODIOT, Department of Mathematics, Facultés Universitaires de Namur, Namur, Belgium

CONFERENCE PARTICIPANTS

W. Alt
Universität Bayreuth
Bayreuth, W. Germay

M. Atteia
Université Paul Sabatier
Toulouse, France

A. Auslender
Université de Clermont-Ferrand II
Aubière, France

J. Baumeister
Frankfurt, W. Germany

H.-P. Blatt
Katolische Universität Eichstätt
Eichstätt, W. Germany

D. Braess
Ruhr-Universität Bochum
Bochum, W. Germany

M. Brokate
Universität Augsburg
Augsburg, W. Germany

J.-C. Cantinat
Université de Dijon
Dijon, France

E. Cavazzuti
Universita di Modena
Modena, Italy

P. Cinquin
Centre Hospitalier Régional de Dijon
Dijon, France

B. Cornet
E.N.S.A.E.
Paris, France

G. Courtade
Institut d'Etudes Juridiques
et Economiques
Bayonne, France

J.-P. Crouzeix
Université de Clermont-Ferrand II
Aubière, France

J.-P. Dedieu
Université Paul Sabatier
Toulouse, France

J.-P. Delahaye
Université des Sciences et Techniques de Lille
Villeneuve d'Ascq, France

J. Denel
Université des Sciences et Techniques de Lille
Villeneuve d'Ascq, France

S. Dolecki
Polish Academy of Sciences
Varsaw, Poland

R. Durier
Université de Dijon
Dijon, France

U. Eckhardt
Universität Hamburg
Hamburg, W. Germany

A. El Qortobi
Université Mohammed V
Oujda, Maroc

A. Fougères
Université de Perpignan
Perpignan, France

D. Gabay
Université P. et M. Curie
Paris, France

S. Gautier
Université de Pau
Pau, France

J. Gauvin
Ecole Polytechnique de Montréal
Montréal, Canada

E. Gekeler
Stuttgart, W. Germany

E. Giner
Université Paul Sabatier
Toulouse, France

B. Gollan
Universität Würzburg
Würzburg, W. Germany

C. Cros
Université d'Avignon
Avignon, France

J. Gwinner
Esslingen, W. Germany

A. Hassouni
Université Paul Sabatier
Toulouse, France

J. Herskovitz
COPPE / UFRJ
Rio de Janeiro, Brazil

J.-B. Hiriart-Urruty
Université Paul Sabatier
Toulouse, France

P. Huard
Direction des Etudes et Recherches d'E.D.F.
Clamart, France

C. King
Universität Stuttgart
Stuttgart, W. Germany

H.-J. Kornstaedt
Freie Universität Berlin
Berlin, W. Germany

W. Krabs
Technische Hochschule Darmstadt
Darmstadt, W. Germany

G. Lebourg
Université de Paris IX - Dauphine
Paris, France

S.-H. Lehnigk
European Research Office of United States Army
London, England

B. Lemaire
Université des Sciences et Techniques du Languedoc
Montpellier, France

C. Lemaréchal
I.N.R.I.A.
Le Chesnay, France

P. Loridan
Université de Dijon
Dijon, France

U. Mackenroth
Universität Bayreuth
Bayreuth, W. Germany

C. Malivert
Université de Limoges
Limoges, France

J.E. Martinez-Legaz
Universidad de Barcelona
Barcelona, Spain

Ph. Michel
Université de Paris I
Paris, France

F. Mignanego
Universita di Genova
Genova, Italy

J. Morgan
Universita di Napoli
Napoli, Italy

W.H. Nguyen
Facultés Universitaires de Namur
Namur, Belgium

J. Noailles
E.N.S.E.E.I.H.T.
Toulouse, France

W. Oettli
Universität Mannheim
Mannheim, W. Germany

N. Ottavy
E.N.S.M.A.
Poitiers, France

D. Pallaschke
Universität Karlsruhe
Karlsruhe, W. Germany

G. Pierra
E.N.S.M.A.
Poitiers, France

G. Pieri
Universita di Genova
Genova, Italy

J.-P. Penot
Université de Pau,
Pau, France

J.-C. Pomerol
Université P. et M. Curie
Paris, France

M. Pradel
Université Paul Sabatier
Toulouse, France

K. Ritter
Universität Stuttgart
Stuttgart, W. Germany

B. Rousselet
Université de Nice
Nice, France

M. Schäl
Universität Bonn
Bonn, W. Germany

K. Schittkowski
Universität Würzburg
Würzburg, W. Germany

I. Singer
I.N.C.R.E.S.T.
Bucuresti, Romania

J. Stoer
Universität Würzburg
Würzburg, W. Germany

J.-J. Strodiot
Facultés Universitaires de Namur
Namur, Belgium

M. Thera
Université de Limoges
Limoges, France

L. Thibault
Université de Pau
Pau, France

J. Vangeldere
Université de Liège
Liège, France

R. Vath
Universität Bayreuth
Bayreuth, W. Germany

M. Volle
Université de Limoges
Limoges, France

J.-C. Yakoubsohn
Université Paul Sabatier
Toulouse, France

U. Zimmermann
Universität Köln
Köln, W. Germany

J. Zowe
Universität Bayreuth
Bayreuth, W. Germany

I. THEORY

HYPERTANGENT CONES FOR A SPECIAL CLASS OF SETS

Szymon Dolecki*

Polish Academy of Sciences
Warsaw, Poland

I. HYPERTANGENT CONES

In recent years much attention has been polarized by the generalized derivative of Clarke [1]. A convenient and insightful approach to this concept (as proposed by Hiriart-Urruty [3] and Rockafellar [5]) is via certain approximating cones to the epigraphs of considered functions. These cones are the following lower limits of the homothety relation (see [2]).

Let X be a linear space in which we consider topologies τ and σ not necessarily compatible with the linear structure of X.

The hypertangent of a subset C of X at x (with respect to τ and σ)

$$\mathscr{C}_C(x) = \mathscr{C}_C^{\tau/\sigma}(x) \tag{1}$$

is composed of those vectors h in X for which for every neighborhood $Q \in \mathscr{N}_\tau(h)$ there exist $t_o > 0$ and $W \in \mathscr{N}_\sigma(x) \cap C$ (= the σ - neighborhood filter of x restricted to the set C) such that

$$Q \cap \frac{1}{t}(C - x') \neq \phi \text{ as } t \leq t_o \text{ , } x' \in W.$$

I adopted the name "hypertangent" for the reasons given in [2]. Other names

*Current affiliation: Dipartimento di Matematica, Università di Trento, Trento Italy

are more widely used, e.g., Rockafellar [5] calls $\mathcal{C}_C^{\tau/\tau}$ the "tangent" and $\mathcal{C}_C^{\iota/\tau}$ the "hypertangent", where ι is the discrete topology. Penot calls $\mathcal{C}_C^{\tau/\tau}$ the "peritangent".
If the topology τ is discrete, we call (1) the radial hypertangent of C at x.

It is known [5] that if $\tau = \sigma$ is a topology compatible with the linear structure and if the interior of the radial hypertangent is nonempty, then the closure of the radial hypertangent is equal to the hypertangent. We shall prove it later.

PROPOSITION 1. Let (X, σ) be a topological vector space and let τ be a translation - stable topology in X. The hypertangent and the radial hypertangent of a τ - closed cone at its vertex are equal.

Under some assumptions on τ and σ, hypertangent cones are convex (no matter what C is) and "extremality-preserving", hence have been studied purposefully as tools in optimality theory. By its nature $\mathcal{C}_C(x)$ must not get off C - x significantly (in this consists the preservation of extremality), but may result disproportionally small with respect to C (always in the vicinity of x). In other words, the hypertangent of a relation (associated with a minimization problem) provides a necessary optimality condition which, however, may be very far from sufficiency. In this context it is important to have a feeling how well the hypertangent cone approximates a given set. With this aim in mind we investigate in this paper hypertangent cones of a special class of sets : cones (see Proposition 1 above).

In this way we acquire useful information about the nature of general hypertangents.

It is significant that "classical" cones (tangent, contingent, radial, etc.) of closed cone at its vertex are equal to this cone. Therefore, this study helps to comprehend the difference between classical approximating cones and hypertangents.

LEMMA 1. Let C be a cone with vertex at 0 in a topological vector space (X, σ) and let τ be a topology in X. A vector h belongs to $\mathcal{C}_C^{\tau/\sigma}(0)$ if and only if for every $Q \in \mathcal{N}_\tau(h)$ and each x in C, $(x + Q) \cap C \neq \phi$. Consequently, an h is in $\mathcal{C}_C^{\iota/\sigma}(0)$, ι denoting the discrete topology, if and only if $h + x \in C$ for every x in C.

Proof. By the definition, h is in $\mathcal{C}_C^{\tau/\sigma}(0)$, whenever for every Q in $\mathcal{N}_\tau(h)$ there are W in $\mathcal{N}_\sigma(0)$ and t_0 such that $(\frac{1}{t}x + Q) \cap C \neq \phi$, (because C is a cone with vertex 0) for x in $W \cap C$ and $0 < t \leq t_0$. Since W is absorbing (radial), every y in C is of the form $\frac{1}{t}x$ with x and t as above ; the proof is completed. □

Proof of Proposition 1. Let h be in the hypertangent of C at 0 and let x be in C. By Lemma 1, every neighborhood Q of h meets C − x ; being closed the latter contains h. By Lemma 1, h lies in the radial hypertangent of C at 0. □

Consider an arbitrary subset D of a vector space X. The radial subset of D from x (denoted $r_D(x)$) is the set of those z for which the interval [x, z] is included in D. Thus $z \in r_D(x)$ iff $x \in r_D(z)$.

The stage of D is composed of those x for which [x, z] is a subset of D for every z in D. We denote it by : stg D.

Therefore, we may write

$$\text{stg } D = \bigcap_{x \in D} r_D(x) = \{h : r_D(h) = D\}. \tag{2}$$

PROPOSITION 2. If C is a cone with vertex at 0, then its radial hypertangent at 0 is equal to the stage of C.

Proof. If h is in stg C, then for every $x \in C$, $\frac{x+h}{2}$ is in C which implies that x + h belongs to the cone C ; h is in $\mathcal{C}_C^{\iota}(0)$. Vice versa, if for every x in C, $h + x \in C$, then for each $x \in C$, $0 < t \leq 1$, we have $h + \frac{(1-t)}{t} x \in C$, or equivalently $th + (1-t)x \in C$, proving that h belongs to stg C. □

So far, for cones (with the vertex at 0), we have reduced the notion of hypertangent cone $\mathcal{C}_C^{\iota/\sigma}(0)$ to a simpler, purely algebraic, concept of stage (stg C). Now we are going to represent $\mathcal{C}_C^{\iota/\sigma}(0)$ in terms of the stage of an arbitrary base of C. We guess that in R^n a similar approach replacing bases by the intersection of C with the unit sphere (and exploiting concepts of stage relative to the geometry of the sphere) might be feasible.

LEMMA 3. Let (X, σ) be a topological vector space. A vector h is in $\mathcal{C}_C^{\iota/\sigma}(0)$ (for a cone with vertex at 0), if and only if for every x in C there is a $m(x) > 0$ such that the interval $[h, m(x)\ x] \subset C$.

Proof. The necessity follows form Proposition 2. Set h, x and m(x) as above. Our objective is to find $0 < t < 1$ and $k > 0$ such that

$$k(th + (1-t)\ m(x)\ x) = h + x$$

to which a solution is $t = \frac{m(x)}{1 + m(x)}$ and $k = \frac{1 + m(x)}{m(x)}$. □

Lef f be a linear form on X such that for every nonzero x in a cone C (with vertex 0) we have $f(x) > 0$. Then the set

$$C_f = C \cup \{x : f(x) = 1\}$$

is called a base of C corresponding to f. It is clear that a base of C exists, whenever C is strictly acute (there is a linear form f strictly positive on $C \setminus \{0\}$).

PROPOSITION 4. The hypertangent cone $\mathcal{C}_C^{\iota/\sigma}(0)$ of a cone C (vertex 0) is equal to the conical hull (about 0) of the stage of every base of C :

$$\text{Cone}(\text{stg } C_f) = \mathcal{C}_C^{\iota/\sigma}(0).$$

where the conical hull of A, Cone (A) $= \cup_{t \geq 0}\ t\,A$.

Proof. $\mathcal{C}_C^{\iota/\sigma}(0)$ is clearly a subset of the conical hull of stg C_f, since for every h in $\mathcal{C}_C^{\iota/\sigma}(0)$ and x in C with $f(x) = f(h) = 1$, $[h, x]$ is a subset of C, hence of C_f. Now it is enough to show that stg C_f is included in $\mathcal{C}_C^{\iota/\sigma}(0)$. Indeed, if h is in the stage of C_f, then for every $x \neq 0$ in C the interval $[h, \frac{x}{f(x)}]$ is included in C_f thus in C. Put $m(x) = f(x)^{-1}$ in Lemma 3. □

The above result allows us to concentrate on objects that prove usually more intuitive and to reduce dimensionality by 1, what turns out to be very useful in euclidean spaces.

II. STAGES

It is an immediate consequence of the definition that a set D is convex, if and only if stg D = D. Moreover

PROPOSITION 5. Every stage is convex.

Proof. Suppose x, y are in stg D, but for some $0 < p < 1$, $px + (1-p)y$ is not in stg D. Therefore, there exists w in D and $0 < r < 1$ such that $v = r(px + (1-p)y) + (1-r)\, w \notin D$. Since y is in stg D, for each $t \geq 1$, $z = tv + (1-t)\, y \notin D$. But on the other hand, for $t = \frac{1}{1 + r(p-1)}$ we have that $z = sx + (1-s)\, w$, where $s = \frac{pr}{pr + (1-r)}$. Because x is in stg D, z must be in D, which is a contradiction. □

Heuristicallly it may be argued that the stage of D is sensible to "non convexity" points of D, because they are responsible for the limitations of visibility within D. In order to formalize this idea we say that an x is internal of D relative E ($x \in \text{int}_E D$), if for every y in E there is t_o such that $(1-t)\, x + ty$ belongs to D as $0 < t < t_o$. Classical internal points are these in $\text{int}_X D$ (algebraic interior of D). Recall that the algebraic closure $\text{cl}_X D$ is the complement of $\text{int}_X(X \setminus D)$ and the points in $\text{bd}_X D = \text{cl}_X D \setminus \text{int}_X D$ are called bounding points of D.

Let D be algebraically closed ($D = \text{cl}_X D$), Let E be a subset of X. We say that a point $x \in D$ is an E - bounding point of D if it is not internal of D relative E. Of course, every E - bounding point of D is a bounding point of D.

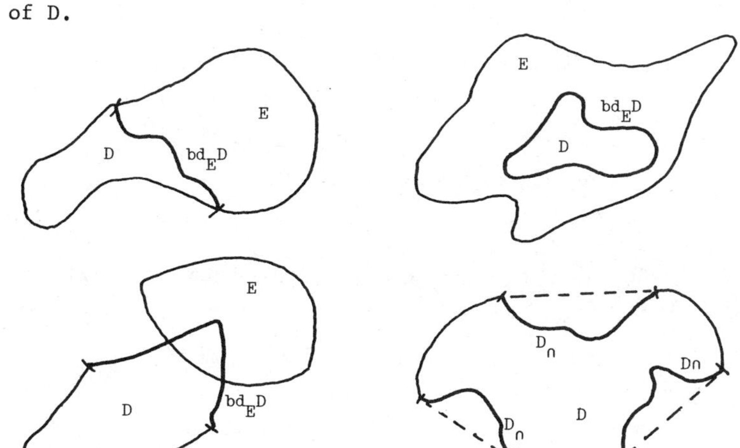

Figure 1. Some examples of E - bounding points of D. The last picture represents the concavity of D. The set of all co D - bounding points of D is called the concavity of D and is denoted by $D_\cap$.

THEOREM 6. Let an algebraically closed set D be included in a convex set E. Then

$$\text{Stg } D = \bigcup_{x \in bd_E D} r_D(x).$$

Proof. In view of (2), we need show that if a vector $h \notin$ stg D, then for some $y \notin bd_E D$, h is not in $r_D(y)$. So assume that there is an x in D such that $[h, x] \setminus D \neq \phi$, that is, the interval [h, x] meets both D and its complement. Since [h, x] with the usual topology is connected, there is a boundary point y of $D \cup [h, x]$ in h, x : there is a sequence $\{t_n\}_{n \in N}$ convergent to 0 such that $(1-t_n)y + t_n h \in E \setminus D$ for all n, because E is convex. Therefore y is an E - bounding point of D, but [y, h] is not a subset of D, thus $h \notin r_D(y)$. □

COROLLARY. Let D be algebraically closed. Then

$$\text{stg } D = \bigcup_{x \in D_\cap} r_D(x).$$

Consequently, the stage of an algebraically closed set D is the intersection of the radial subsets of D from its bounding points.

Recall that an x is an extremal point of a set D (not necessarily convex), if for every y and z such that $x \in [y, z] \subset D$, we must have that x = y = z. Attempts ot represent the stage as the intersection of radial subsets from extremal points break already in R^3.

Example. Consider the union of segments [(0, 0, 0), (0, 0, 1)], [(0, 0, 0), (0, 1, 0)], [(0, 0, 1), (0, 1, 0)] and the ball B((0, 0, 3), 1). Take the union of all segments joining points of the ball with those from the boundary of the triangle. So constructed set has the property that

$$\bigcap_{x \in \text{Ext} D} r_D(x) \setminus \text{stg } D = \{(0,0,0), (0,0,1), (0,1,0)\}.$$

Let K be a convex subset of D containing x. Then obviously $K \subset r_D(x)$. Therefore representations of the stage as the intersection of convex subsets of D give a finer description of the stage than the analogous ones that use radial subsets.

Let $\mathfrak{M}_D(x)$ denote the family of maximal convex subsets of D which contain x. We observe that for every x in D, the stage is a subset of each K in $\mathfrak{M}_D(x)$:

$$\text{stg } D \subset \bigcap_{K \in \mathfrak{M}_D(x)} K, \tag{3}$$

because Co({x}, stg D) ⊂ D.

THEOREM 7. Let D be an algebraically closed subset of a convex set E. Then

$$\text{stg } D = \bigcap_{x \in bd_E D} , \bigcap_{K \in \mathfrak{M}_D(x)} K.$$

Proof. By (3) we have that for every $x \in bd_E D$,
$\text{stg } D \subset \bigcap_{K \in \mathfrak{M}_D(x)} K \subset r_D(x)$. Thus

$$\text{stg } D \subset \bigcap_{x \in bd_E D} \bigcap_{K \in \mathfrak{M}_D(x)} K \subset \bigcap_{x \in bd_E D} r_D(x)$$

hence by Theorem 6, the proof is complete. □

COROLLARY. For an algebraically closed set D

$$\text{stg } D = \bigcap_{x \in D_\cap} \bigcap_{K \in \mathfrak{M}_D(x)} K.$$

The operation stg : $2^X \to 2^X$ is not isotone, thus not a relation. This is easily seen by taking the union of two convex sets being non convex. It is clear that if the stage is not empty, then the set must be connected and moreover must not contain holes.

A conjecture is that the stage of a compact subset of a locally convex topological vector space is the intersection of maximal convex sets that meet extremal points that lie in the concavity.

We terminate by drawing some sets and indicating (by stripped area) their stages :

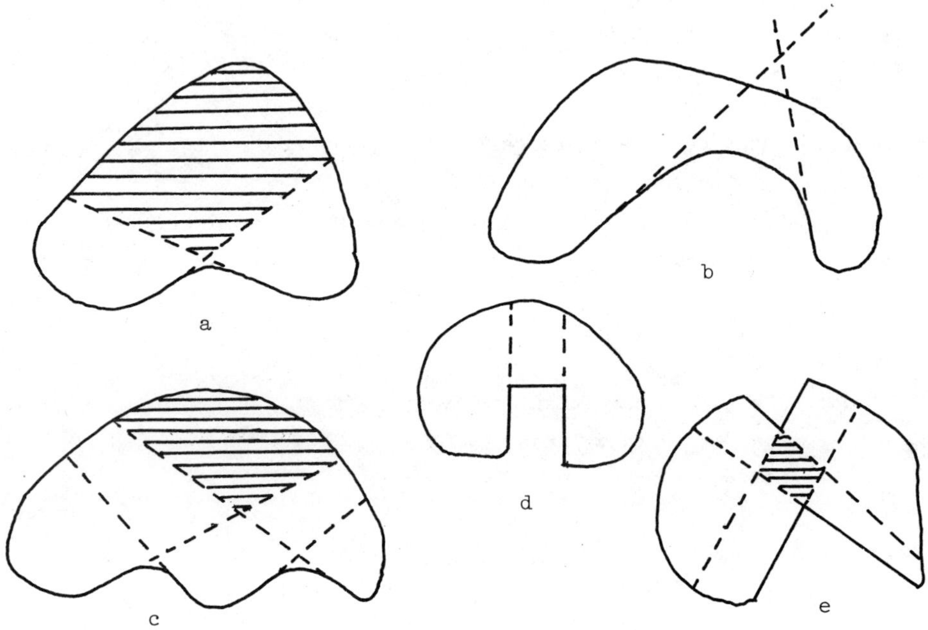

Figure 2. Stages of flat figures. In b) and d) we easily find two disjoint maximal convex subsets.

Acknowledgement : My thanks go to the referee for several suggestions of improvements.

REFERENCES

1. F.H. Clarke, Generalized gradients and applications, Trans. A.M.S. 205 (1975), 247-262.

2. S. Dolecki, Tangency and differentiation : some applications of convergence theory, to appear.

3. J.-B. Hiriart-Urruty, Contributions à la programmation mathématique : cas déterministe et stochastique, Thèse, Université de Clermont-Ferrand II (1977).

4. J.-P. Penot, A characterization of tangential regularity, Nonlinear Analysis, Theory, Methods Appl. 5 (1981), 625-663.

5. R.-T. Rockafellar, Generalized directional derivatives and subgradients of nonconvex functions, Can. J. Math. 32 (1980), 257-280.

OPTIMIZATION BY LEVEL SET METHODS
I : DUALITY FORMULAE

Ivan Singer

National Institute for Scientific and Technical Creation
Bucharest, Romania

Let G be a non-empty subset of a (real) locally convex space F and $h : F \to \overline{R}$ a functional. We show that the conditions of existence of functionals $\psi \in F^*$, $\psi \neq 0$, which support G and separate G from certain level sets of h, closedly or openly or nicely (in the sense of V. Klee [7]) imply duality formulae which reduce the computation of inf h(G) to the computation of the infima of h on some closed half-spaces or strips or closed strips containing G and that, conversely, under some additional connectedness assumptions on G and certain level sets of h, these support and separation conditions are also necessary for our duality formulae to hold.

INTRODUCTION

In the present paper we shall study the following general optimization problem : Given a (real) locally convex space F, a subset G of F (which will be assumed non-empty throughout the paper), called the constraint set, and a functional $h : F \to \overline{R} = [-\infty, +\infty]$, find convenient formulae for

$$\inf h(G) = \inf_{g \in G} h(g). \tag{0.1}$$

We shall not assume the existence of "solutions" for (0.1), i.e., of elements $g_o \in G$ such that

$$h(g_o) = \inf h(G) \; ; \tag{0.2}$$

we intend to study elsewhere the solutions g_o, by methods similar to those of the present paper.

Our aim is to obtain duality formulae for (0.1), involving functionals $\psi \in F^*$ (the set of all continuous linear functionals on F). The usefulness of such results for applications is due to the fact that in various concrete spaces F the general form of continuous linear functionals $\psi \in F^*$ is well known and simple. More precisely, we want to express inf h(G) with the aid of inf h(M), for some suitable sets M related to G, defined by means of functionals $\psi \in F^*$, $\psi \neq 0$, such as closed half-spaces containing G or strips containing G, which present, in some cases, the advantage that there exist simple formulae for the computation of inf h(M).

In the present paper, in order to study the general optimization problem (0.1), we shall not use epigraph methods, but only level set methods, and we shall make no convexity assumptions on G or on the level sets of h. Our main tool will be the separation of G from certain level sets of h by functionals $\psi \in F^*$, $\psi \neq 0$. We shall show that the "most suitable" kinds of separation to be used, in the sense that they yield conditions which are sufficient and, in many cases, also necessary, for certain duality formulae to hold, are the "closed", "open", and "nice" separation in the sense of V. Klee [7]. More precisely, we shall show that the conditions of existence of functionals $\psi \in F^*$, $\psi \neq 0$ (or, equivalently, of hyperplanes) which support G and separate G from certain level sets of h, in one of the above senses, imply duality formulae which reduce the computation of inf h(G) to the computation of the infima of h on some closed half-spaces or strips or closed strips containing G and that, conversely, under some additional connectedness assumptions on G and certain level sets of h, these support and separation conditions are also necessary for our duality formulae to hold. In §1 we shall study theorems of weak duality, which express inf h(G) as the supremum, over certain sets of functionals $\psi \in F^*$, of the infima of h over some closed half-spaces or strips or closed strips containing G, defined with the aid of the functionals ψ, but the supremum need not be attained. In §2 we shall consider theorems of strong duality, which express inf h(G) as above, but with the supremum being attained.

Throughout the paper we shall adopt the usual conventions $\inf \phi = +\infty$, $\sup \phi = -\infty$, where ϕ denotes the empty set.

Let us mention that a generalization of closed separation in the sense of V. Klee [7] has been used by G.Š. Rubinštein ([14], [15]), who has developed a very general scheme of obtaining duality theorems for optimization problems ; however, Rubinštein did not obtain duality theorems expressing $\inf h(G)$ with the aid of $\inf h(M)$, with suitable sets M related to G. Also, we mention that our use of support functionals ψ of G (instead of general functionals $\psi \in F^*$, $\psi \neq 0$), which separate G form certain level sets of h, has the advantage that support functionals can be easily computed for certain classes of sets G, as we shall show.

The results of the present paper imply some of the previous results giving sufficient conditions for certain duality formulae to hold obtained by level set methods or other methods, e.g. in [17] and in the references thereof. Moreover, some of our results lead to new facts for the particular case of quasi-convex optimization and some even for convex optimization.

Finally, we thank Professors J.-P. Crouzeix, P.-J. Laurent and W. Oettli for indicating to us useful references and for calling to our attention the duality results in terms of half-spaces, mentioned in remarks 1.4 b) and 2.3 a).

I. THEOREMS OF WEAK DUALITY

I.0. PRELIMINARIES. We say that a (closed) hyperplane

$$H = \{y \in F \mid \psi(y) = c\}, \tag{1.1}$$

where $\psi \in F^*$, $\psi \neq 0$, $c \in R$, supports a set $G \subset F$ (or, that H is a support hyperplane of G), if G is contained in one of the two closed half-spaces $\{y \in F \mid \psi(y) \leq c\}$ or $\{y \in F \mid \psi(y) \geq c\}$ determined by H, but for each $\varepsilon > 0$, G is not contained in any one of the closed half-spaces $\{y \in F \mid \psi(y) \leq c-\varepsilon\}$, $\{y \in F \mid \psi(y) \geq c + \varepsilon\}$ (however, we do not require that $G \cap H \neq \phi$). Clearly, this happens if and only if either $c = \sup \psi(G)$ or $c = \inf \psi(G)$; thus, every hyperplane H supporting G can be written in the form

$$H_{\psi,G} = \{y \in F \mid \psi(y) = \sup \psi(G)\}, \tag{1.2}$$

where $\psi \in F^*$, $\psi \neq 0$, $\sup \psi(G) < +\infty$, and conversely, every hyperplane of the

form (1.2), where $\psi \in F^*$, $\psi \neq 0$, sup $\psi(G) < +\infty$, supports G.
We say that a functional $\psi \in F^*$ supports a set G (or, that ψ is a support functional of G), if $\psi \neq 0$ and sup $\psi(G) < +\infty$. This terminology is justified by (1.2) ; however, let us note that if inf $\psi(G) > -\infty$, then $-\psi$ is a support functional of G, but ψ need not be a support functional of G.
For any set $G \subset F$ we shall denote by G^s the set of all support functionals of G, i.e.,

$$G^s = \{\psi \in F^* \mid \psi \neq 0, \sup \psi(G) < +\infty\} ; \tag{1.3}$$

thus, G^s is nothing else than $G^s = \operatorname{dom} \chi_G^* \setminus \{0\}$, where χ_G^* is the (convex) conjugate of the indicator functional of G [13]. Let us mention that the (generally, non-linear) functional χ_G is sometimes also called in the literature "the support functional of G" (see e.g. [13], pp. 28, 112), but we shall not use this terminology.
In the sequel we shall use various types of non-symmetric separation properties, defined by adding to the usual separation and to the types of separation of [7], a condition for "orientation" of the half-spaces, as follows : we shall say that a hyperplane (1.1) :

a) separates G_1 from G_2, if $G_1 \subset \{y \in F \mid \psi(y) \leq c\}$, $G_2 \subset \{y \in F \mid \psi(y) \geq c\}$;

b) closely separates G_1 from G_2, if $G_1 \subset \{y \in F \mid \psi(y) \leq c\}$, $G_2 \subset \{y \in F \mid \psi(y) > c\}$;

c) Openly separates G_1 from G_2, if $G_1 \subset \{y \in F \mid \psi(y) < c\}$, $G_2 \subset \{y \in F \mid \psi(y) \geq c\}$;

d) nicely separates G_1 form G_2, if either b) or c) holds. Correspondingly, we shall say that a functional $\psi \in F^* \setminus \{0\}$ separates, closedly separates, etc. G_1 from G_2, if there exists $c \in R$ such that the hyperplane (1.1) has the corresponding separation property.
Given a set $G \subset F$ and a functional $h : F \to \overline{R}$, in the present section an important role will be played by the level sets

$$A_c = \{y \in F \mid h(y) < c\} \qquad (c < \inf h(G) = a), \tag{1.4}$$

$$S_c = \{y \in F \mid h(y) \leq c\} \qquad (c < a). \tag{1.5}$$

I.1. RESULTS OF WEAK DUALITY IN TERMS OF CLOSED HALF-SPACES CONTAINING G

PROPOSITION 1.1. Let F be a locally convex space, G a subset of F, $h : F \to \overline{R}$ a functional, $c \in R$ and $\psi_c \in F^*$.

i) We have

$$\inf_{\substack{y\in F \\ \psi_c(y)\le\sup\psi_c(G)}} h(y) \ge c \tag{1.6}$$

if and only if

$$\sup \psi_c(G) < \psi_c(y) \qquad (y \in A_c). \tag{1.7}$$

i') We have

$$\inf_{\substack{y\in F \\ \psi_c(y)\ge\inf\psi_c(G)}} h(y) \ge c \tag{1.6'}$$

if and only if

$$\psi_c(y) < \inf \psi_c(G) \qquad (y \in A_c) ; \tag{1.7'}$$

when $A_c = \phi$, we consider that (1.7) and (1.7') are vacuously satisfied.

Proof. i) If (1.6) does not hold, then there exists $y_o \in F$ with $\psi_c(y_o) \le \sup \psi_c(G)$, such that $h(y_o) < c$ (thus, $y_o \in A_c$), so (1.7) is not satisfied for $y = y_o$.
Conversely, if (1.7) is not satisfied for some $y = y_o \in A_c$, then $\psi_c(y_o) \le \sup \psi_c(G)$ and $h(y_o) < c$, so (1.6) does not hold.
Finally, i') is equivalent to i), considering $\psi'_c = -\psi_c$. This completes the proof of Proposition 1.1. □

Remark 1.1. For each $\psi \in F^*$, $\psi \neq 0$, let us call closed half-space in the direction ψ any set of the form

$$D = \{y \in F \mid \psi(y) \le \beta\} = \psi^{-1}((-\infty, \beta]), \tag{1.8}$$

where $\beta \in \overline{R}$ (when $\beta = +\infty$ or $\beta = -\infty$, we have $D = F$ or $D = \phi$ respectively). Then, denoting by $D_{\psi,G}$ the smallest closed half-space in the direction ψ, containing G, Proposition 1.1 i) means, geometrically, that we have

$$\inf_{y \in D_{\psi_c,G}} h(y) \ge c \tag{1.9}$$

if and only if ψ_c separates closedly G from A_c.

COROLLARY 1.1. Let F, G, h, c and ψ_c be as in Proposition 1.1.

i) If

$$\sup \psi_c(G) < \psi_c(y) \qquad (y \in S_c), \tag{1.10}$$

then we have (1.6).

ii) If

$$\inf_{\substack{y \in F \\ \psi_c(y) \leq \sup \psi_c(G)}} h(y) > c, \tag{1.11}$$

then (1.10) holds.

i') If

$$\psi_c(y) < \inf \psi_c(G) \qquad (y \in S_c), \tag{1.10'}$$

tehn we have (1.6').

ii') If

$$\inf_{\substack{y \in F \\ \psi_c(y) \geq \inf \psi_c(G)}} h(y) > c, \tag{1.11'}$$

then (1.10') holds.

Proof. Parts i), i') follow from $A_c \subset S_c$ and the sufficiency parts of Proposition 1.1 i) and i') respectively.
If we have (1.11), then, by the necessity part of Proposition 1.1 i) there holds $\sup \psi_c(G) < \psi_c(y)$ $(y \in A_r)$ for some $r > c$, whence (1.10) follows. This proves ii).
Finally, ii') is equivalent to ii), considering $\psi_c' = -\psi_c$. This completes the proof of Corollary 1.1. □

THEOREM 1.1. Let F be a locally convex space, G a subset of F with $G^s \neq \phi$ and $h : F \to \overline{R}$ a functional. The following statements are equivalent:

1°) For each $c < \inf h(G) = a$ there exists $\psi_c \in G^s$ satisfying (1.7).

2°) For each $c < a$ there exists $\psi_c \in G^s$ satisfying (1.10).

3°) There holds

$$\inf h(G) = \sup_{\psi \in G^s} \inf_{\substack{y \in F \\ \psi(y) \leq \sup \psi(G)}} h(y). \tag{1.12}$$

4°) For each $c < a$ there exists $\psi_c \in -G^s$ satisfying (1.7').

5°) For each $c < a$ there exists $\psi_c \in -G^s$ satisfying (1.10').

6°) There holds

$$\inf h(G) = \sup_{\psi\in -G^s} \inf_{\substack{y\in F \\ \psi(y)\geq \inf \psi(G)}} h(y). \tag{1.12'}$$

Proof. The implication 2°) $\Longrightarrow$ 1°) is obvious (since $A_c \subset S_c$).
1°) $\Longrightarrow$ 3°). We have always (i.e., without any assumptions on G and h) the inequality

$$a = \inf h(G) \geq \sup_{\psi\in G^s} \inf_{\substack{y\in F \\ \psi(y)\leq \sup \psi(G)}} h(y) = b, \tag{1.13}$$

and hence, if $a = -\infty$ (so 1°) is vacuously satisfied), then (1.12) holds. If $-\infty < c < a$, then, by $G^s \neq \phi$ (when $A_c = \phi$) or 1°) (when $A_c \neq \phi$) and by Proposition 1.1 i), there exists $\psi_c \in G^s$ satisfying (1.6), whence $b \geq c$. Consequently, $b \geq \sup_{c<a} c = a$, whence, by (1.13), we obtain (1.12).
3°) $\Longrightarrow$ 2°). If $a = -\infty$ (so (1.12) holds by (1.13)), then 2°) is vacuously satisfied. If $a > -\infty$, then, by (1.12), for each $c < a$ there exists $\psi_c \in G^s$ satisfying (1.11), whence also (1.10) (by Corollary 1.1 ii)).
Finally, considering $\psi'_c = -\psi_c$, $\psi' = -\psi$, we obtain the equivalences $n^o \Longleftrightarrow (n+3)^o$ ($n = 1,2,3$). This completes the proof of Theorem 1.1. □

Remark 1.2. a) The assumption $G^s \neq \phi$ of Theorem 1.1 implies $\overline{G} \neq F$, and, if G is convex, the converse implication is also true.
Indeed, if $\overline{G} = F$, then for each $\psi \in F^*$, $\psi \neq 0$ we have $\sup \psi(G) = \sup \psi(\overline{G}) = +\infty$, so $G^s = \phi$; note also that in this case the right hand side of (1.12) is $-\infty$, while the left hand side of (1.12) may be any $a \in \overline{R}$. Conversely, assume that G is convex and that $\overline{G} \neq F$, and let $y' \in F \setminus \overline{G}$. Then, since $\overline{G}$ is convex, there exists (see e.g. [3], p. 417) a functional $\psi \in F^*$, $\psi \neq 0$, such that $\sup \psi(G) = \sup \psi(\overline{G}) < \psi(y') < +\infty$, so $\psi \in G^s$, which proves our assertion.

b) If $\psi \notin G^s$, then $\{y \in F \mid \psi(y) \leq \sup \psi(G)\} = F$. Consequently, if one (and hence each) of 1°),...,6°) of Theorem 1.1 holds, then we have

$$\inf h(G) = \sup_{\psi\in F^*} \inf_{\substack{y\in F \\ \psi(y)\leq \sup \psi(G)}} h(y) = \sup_{\psi\in G^s\cup\{0\}} \inf_{\substack{y\in F \\ \psi(y)\leq \sup\psi(G)}} h(y), \tag{1.14}$$

and the corresponding equalities for $\sup \psi(G)$ replaced by $\inf \psi(G)$.

c) Assume that one of the following equivalent conditions is satisfied : α) $A_c = \phi$ $(c < a)$; β) $S_c = \phi$ $(c < a)$; γ) $A_a = \phi$; δ) $\inf h(F) = \inf h(G)$. Then, since $G \subset \{y \in F \mid \psi(y) \leq \sup \psi(G)\} \subset F$ and $G \subset \{y \in F \mid \psi(y) \geq \inf \psi(G)\} \subset F$, we have

$$\inf h(G) = \inf_{\substack{y \in F \\ \psi(y) \leq \sup \psi(G)}} h(y) = \inf_{\substack{y \in F \\ \psi(y) \geq \inf \psi(G)}} h(y) \; (\psi \in F^*, \psi \neq 0), \tag{1.15}$$

whence (1.12) and (1.12'), provided that $G^s \neq \phi$. Also, in this case (1.7), (1.10), (1.7'), (1.10') of Theorem 1.1 are satisfied (vacuously) for any $\psi_c \in G^s$ or $\psi_c \in -G^s$ respectively. Thus, Theorem 1.1 presents interest only when α) − δ) above do not hold. A similar remark is also valid for the other duality theorems of this paper.

Remark 1.3. a) Geometrically, formula (1.12) means that

$$\inf h(G) = \sup_{\psi \in G^s} \inf h(D_{\psi,G}), \tag{1.16}$$

where $D_{\psi,G}$ is as in Remark 1.1 ; note that, if $\psi \in G^s$, then $D_{\psi,G}$ is proper (i.e., $D_{\psi,G} \neq F$).

b) Formula (1.12) may be regarded as a formula of "surrogate duality" in the sense of [4] and [5] (see also [10]), i.e., as replacing the constraint set G of the initial optimization problem inf h(G) by the "surrogate constraint sets" $\{y \in F \mid \psi(y) \leq \sup \psi(G)\}$ $(\psi \leq G^s)$. A similar remark is also valid for the other duality formulae of this paper.

COROLLARY 1.2. If $G^s \neq \phi$ and one (and hence each) of 1°),..., 6°) of Theorem 1.1 holds, then

$$\inf h(G) = \sup_{\substack{\psi \in G^s,\ \beta \in \overline{R} \\ \sup \psi(G) \leq \beta}} \inf_{\substack{y \in F \\ \psi(y) \leq \beta}} h(y) \tag{1.17}$$

or, geometrically,

$$\inf h(G) = \sup_{\substack{D \in \mathcal{D} \\ D \supset G}} \inf h(D), \tag{1.18}$$

where $\mathcal{D}$ denotes the collection of all closed half-spaces in F.

Proof. By (1.16) we have the inequality ≤ in (1.18). On the other hand, for any set $D \supset G$ we have $\inf h(G) \geq \inf h(D)$, whence (1.18) follows. □

It will be worth while to state separately the following sufficient condition in order to have (1.12), (1.12'), (1.18) :

THEOREM 1.2. Let F be a locally convex space, G a proper subset of F and $h : F \to \overline{R}$ a functional, such that G and the (possibly empty) level sets S_c $(c < \inf h(G))$ are convex and closed for a locally convex topology τ on F, weaker than or equal to the initial topology on F, and either G or the sets S_c $(c < \inf h(G))$ are compact for τ. Then we have 2°) of Theorem 1.1, whence also (1.12), (1.12'), (1.18).

Proof. Since G is a proper τ-closed subset of F, by our assumption on τ we have $\overline{G} = G \neq F$ (in the initial topology). Hence, since G is convex, by Remark 1.2 a) we get $G^s \neq \phi$. Thus, if there exists no $c < \inf h(G)$ with $S_c \neq \phi$, then 2°) of Theorem 1.1 is satisfied for any $\psi_c \in G^s$. On the other hand, if there exists $c < \inf h(G)$ with $S_c \neq \phi$, then, clearly, $G \cap S_c = \phi$. Also, by our assumptions, G and S_c are proper convex subsets of F, closed for τ, and one of them is compact for τ. Hence, there exists (see e.g. [3], p. 417) a functional $\psi_c \in (F, \tau)^* \subset F^*$, $\psi_c \neq 0$, satisfying $\sup \psi_c(G) < \inf \psi_c(S_c)$, whence (1.10) and $\psi_c \in G^s$. Thus, we have 2°) of Theorem 1.1, whence also (1.12), (1.12'), (1.18), which completes the proof of Theorem 1.2. □

As an example of application, let us give

COROLLARY 1.3. Let F be a reflexive Banach space, G a proper closed convex subset of F and $h : F \to \overline{R}$ a lower semi-continuous quasi-convex functional, such that either G or the level sets S_c $(c < \inf h(G))$ are bounded. Then we have (1.12), (1.12'), (1.18).

Proof. Since G and S_c $(c < a)$ are closed and convex, they are also weakly closed (see e.g. [3], p. 422). Hence, since either G or the sets S_c $(c < a)$ are bounded and since F is a reflexive Banach space, either G or the sets S_c $(c < a)$ are weakly compact. Consequently, we can apply Theorem 1.2 (with τ = the weak topology on F), which completes the proof. □

We shall show now that for some important classes of convex sets G, such as convex cones and linear manifolds, the sets G^s can be computed and have particulary simple expressions. Thus, the use of formulae (1.12) and

(1.12'), involving G^s (instead of (1.14) and the corresponding one with $\inf \psi(G)$), is convenient for applications.
We recall that the polar of a set $G \subset F$ is the set $G^o = \{\psi \in F^* \mid \sup \psi(G) \leq 1\}$.

A) If G is a convex cone with vertex y_o, then

$$G^s \cup \{0\} = (G-y_o)^o = \{\psi \in F^* \mid \sup \psi(G) = \psi(y_o)\}. \tag{1.19}$$

Consequently, formula (1.12') becomes

$$\inf h(G) = \sup_{\substack{0 \neq \psi \in F^* \\ \inf \psi(G) = \psi(y_o)}} \inf_{\substack{y \in F \\ \psi(y) \geq \psi(y_o)}} h(y) \tag{1.20}$$

and formula (1.12) changes similarly.

B) If G is a linear manifold, say $G = y_o + S$, where $y_o \in F$ and S is a linear subspace of F, then,

$$G^s \cup \{0\} = S^\perp = \{\psi \in F^* \mid \psi(s) = 0 \quad (s \in S)\}. \tag{1.21}$$

Consequently, formula (1.12') becomes

$$\inf h(G) = \sup_{0 \neq \psi \in S^\perp} \inf_{\substack{y \in F \\ \psi(y) \geq \psi(y_o)}} h(y) \tag{1.22}$$

and formula (1.12) changes similarly.
If $G = y_o + S$ is a linear manifold, there is only one case in which the assumption of τ-compactness of G in Theorem 1.2 is satisfied, namely, when $G = \{y_o\}$, $S = \{0\}$ (hence $S^\perp = F^*$). In view of the importance for applications (via primal functionals) of the formulae expressing the value of a functional at a point (see [17], [18]), let us state now separately this particular case of Theorem 1.2, as follows :

COROLLARY 1.4. Let F be a locally convex space, $y_o \in F$, and $h : F \to \overline{R}$ a functional, such that the level sets S_c $(c < h(y_o))$ are convex and closed for a locally convex topology τ on F, weaker than or equal to the initial topology on F. Then

$$h(y_o) = \sup_{0 \neq \psi \in F^*} \inf_{\substack{y \in F \\ \psi(y) \geq \psi(y_o)}} h(y). \tag{1.23}$$

Remark 1.4. a) The right-hand side of (1.23) is, essentially, the "normalized second quasi-conjugate" of h at y_o in the sense of [6].

b) In particular, the assumption of Corollary 1.4 is satisfied for all $y_o \in F$ when h is quasi-convex and τ - lower semi-continuous (which, by our assumption on τ, implies that h is also lower semi-continuous for the initial topology on F). For $h : F \to \overline{R}$ quasi-convex and lower semi-continuous for the initial topology, Corollary 1.4 has been obtained by J.-P. Crouzeix ([2], lemma 3). Actually, J.-P. Crouzeix [2] has proved (1.23) under the weaker assumption that h is quasi-convex and lower semi-continuous at y_o. This result can be also deduced from Theorem 1.1, as follows : Let $G = \{y_o\}$ and let $c < a = h(y_o)$. Then, since h is lower semi-continuous at y_o, there exists a convex neighbourhood V of y_o such that $h(y) < c$ $(y \in V)$. Now, if $S_c = \phi$, then 5°) of Theorem 1.1 is satisfied for any $\psi_c \in -G^s = F^* \setminus \{0\}$. On the other hand, if $S_c \neq \phi$, then S_c and V are non-empty convex subsets of F, with Int $V \neq \phi$, $S_c \cap V = \phi$, and hence there exists (see e.g. [3], p. 417, theorem 8) a functional $\psi_c \in F^*$, $\psi_c \neq 0$, such that $\sup \psi_c(S_c) \leq \inf \psi_c(V)$. But, since $y_o \in$ Int V, we have $\inf \psi_c(V) < \psi_c(y_o)$ (by §2, lemma 2.1 below), whence $\sup \psi_c(S_c) < \psi_c(y_o)$. Therefore, we have (1.10'), so 5°) of Theorem 1.1 is again satisfied. Consequently, we have (1.12') with $G = \{y_o\}$, i.e., (1.23), which completes the proof.

I.2. RESULTS OF WEAK DUALITY IN TERMS OF STRIPS CONTAINING G

PROPOSITION 1.2. Let F be a locally convex space, G a subset of F, $h : F \to \overline{R}$ a functional, $c \in R$ and $\psi_c \in F^*$.

i) If one of the four conditions (1.7),

$$\psi_c(g) < \inf \psi_c(A_c) \qquad (g \in G), \tag{1.24}$$

(1.7') or

$$\sup \psi_c(A_c) < \psi_c(g) \qquad (g \in G), \tag{1.24'}$$

is satisfied (if $A_c = \phi$, then (1.24) and (1.24') hold and we consider that (1.7) and (1.7') are vacuously satisfied), then

$$\inf_{\substack{y \in F \\ \psi_c(y) \in \psi_c(G)}} h(y) \geq c. \tag{1.25}$$

ii) Conversely, if $\psi_c(G)$ and $\psi_c(A_c)$ are intervals in R and if (1.25) holds, then ψ_c satisfies one of (1.7), (1.24), (1.7') or (1.24').

Proof. i) If (1.25) does not hold, then there exists $y_o \in F$ with $\psi_c(y_o) \in \psi_c(G)$, such that $h(y_o) < c$ (thus, $y_o \in A_c$). Then

$$\inf \psi_c(G) \leq \psi_c(y_o) \leq \sup \psi_c(G),$$

so none of (1.7) or (1.7') holds. Also, $\psi_c(y_o) = \psi_c(g_o)$ for some $g_o \in G$, whence

$$\inf \psi_c(A_c) \leq \psi_c(y_o) = \psi_c(g_o) \leq \sup \psi_c(A_c)$$

so none of (1.24) or (1.24') holds.

ii) Assume that $\psi_c(G)$ and $\psi_c(A_c)$ are intervals in R and that we have (1.25). Then, by (1.25) and $\psi_c \neq 0$,

$$A_c \subset \{y \in F \mid \psi_c(y) \notin \psi_c(G)\}, \tag{1.26}$$

whence, since $\psi_c(G)$ is an interval in R, we obtain

$$A_c \subset \{y \in F \mid \psi_c(y) \geq \sup \psi_c(G)\} \cup \{y \in F \mid \psi_c(y) \leq \inf \psi_c(G)\}. \tag{1.27}$$

We claim that A_c lies in one of the two closed half-spaces occurring in (1.27). Indeed, assume, a contrario, that there exist $y_1, y_2 \in A_c$, $y_1 \neq y_2$, such that

$$\psi_c(y_1) \geq \sup \psi_c(G) \geq \inf \psi_c(G) \geq \psi_c(y_2). \tag{1.28}$$

Then, by (1.26), $\psi_c(y_1) \neq \psi_c(y_2)$. Furthermore, since $\psi_c(A_c)$ is an interval in R, by (1.28) and $\psi_c(y_1), \psi_c(y_2) \in \psi_c(A_c)$ we obtain

$$\psi_c(G) \subset [\psi_c(y_2), \psi_c(y_1)] \subset \psi_c(A_c), \tag{1.29}$$

which contradicts (1.26). This proves the claim. Thus, we have either

$$\sup \psi_c(G) \leq \inf \psi_c(A_c), \tag{1.30}$$

or

$$\sup \psi_c(A_c) \leq \inf \psi_c(G). \tag{1.30'}$$

Assume now, a contrario, that none of (1.7), (1.24), (1.7') or (1.24') holds and that we are in case (1.30) (the case of (1.30') is similar).

Then, by (1.30) and non-(1.7), there exists $y_o \in A_c$ such that $\psi_c(y_o) = \inf \psi_c(A_c)$ and, similarly, by (1.30) and non-(1.24), there exists $g_o \in G$ such that $\psi_c(g_o) = \inf \psi_c(A_c)$. But, then $y_o \in A_c$ and

$$\psi_c(y_o) = \inf \psi_c(A_c) = \psi_c(g_o) \in \psi_c(G), \tag{1.31}$$

in contradiction with (1.26). This completes the proof of Proposition 1.2. □

Remark 1.5. a) For each $\psi \in F^*$, $\psi \neq 0$, let us call strip generated by ψ any set of the form $B = \psi^{-1}(I)$, where $I \subset R$ is connected, i.e.,

$$B = \{y \in F \mid \psi(y) \in <\alpha, \beta>\} = \psi^{-1}(<\alpha, \beta>), \tag{1.32}$$

where $\alpha, \beta \in \overline{R}$, $\alpha \leq \beta$, and where $<\alpha, \beta>$ denotes either one of the intervals $[\alpha, \beta]$, $(\alpha, \beta]$, $[\alpha, \beta)$ or (α, β) (when $\beta = +\infty$, $\alpha = -\infty$, we have $B = F$; when either $\beta = -\infty$ or $\alpha = +\infty$, we have $B = \phi$). Thus, B is either the portion of F "between" the hyperplanes $H_{\psi,\alpha} = \{y \in F \mid \psi(y) = \alpha\}$ and $H_{\psi,\beta}$ (when $-\infty < \alpha \leq \beta < +\infty$), with one or both of $H_{\psi,\alpha}$, $H_{\psi,\beta}$ included or excluded, or a half-space. In particular, if G is a subset of F such that $\psi(G)$ is an interval in R, then

$$\{y \in F \mid \psi(y) \in \psi(G)\} = B_{\psi,G}, \tag{1.33}$$

where $B_{\psi,G}$ denotes the smallest strip generated by ψ, containing G. Indeed, since $\psi(G)$ is an interval, $\{y \in F \mid \psi(y) \in \psi(G)\}$ is a strip containing G, whence $B_{\psi,G} \subset \{y \in F \mid \psi(y) \in \psi(G)\}$. Conversely, if $B_{\psi,G} = \{y \in F \mid \psi(y) \in <\alpha, \beta>\}$ and if $y_o \in F$, $\psi(y_o) \in \psi(G)$, so $\psi(y_o) = \psi(g_o)$ for some $g_o \in G$, then, by $G \subset B_{\psi,G}$ we obtain $\psi(y_o) = \psi(g_o) \in <\alpha, \beta>$, whence $y_o \in B_{\psi,G}$; thus, $\{y_o \in F \mid \psi(y_o) \in \psi(G)\} \subset B_{\psi,G}$, which proves (1.32). Consequently, if $\psi_c(G)$ and $\psi_c(A_c)$ are intervals in R, then Proposition 1.2 means, geometrically, that we have

$$\inf_{y \in B_{\psi_c,G}} h(y) \geq c \tag{1.34}$$

if and only if either ψ_c or $-\psi_c$ separates nicely G from A_c.

b) The assumptions that $\psi_c(G)$ and $\psi_c(A_c)$ are intervals in R cannot be omitted in Proposition 1.2, as shown by simple examples.

PROPOSITION 1.3. Let F, G, h, c and ψ_c be as in Proposition 1.2.

i) If one of the four conditions (1.10),

$$\psi_c(g) < \inf \psi_c(S_c) \qquad (g \in G), \tag{1.35}$$

(1.10') or

$$\sup \psi_c(S_c) < \psi_c(g) \qquad (g \in G) \tag{1.35'}$$

is satisfied, then we have (1.25).

ii) If $\psi_c(G)$ and $\psi_c(S_c)$ are intervals in R and if

$$\inf_{\substack{y \in F \\ \psi_c(y) \in \psi_c(G)}} h(y) > c, \tag{1.36}$$

then ψ_c satisfies one of (1.10), (1.35), (1.10') or (1.35').

Proof. Part i) follows from $A_c \subset S_c$ and Proposition 1.2 i). The proof of part ii) is similar to that of Proposition 1.2 ii), replacing A_c by S_c. □

In the sequel we shall use the well-known fact that connectedness is preserved under continuous mappings.

THEOREM 1.3. Let F be a locally convex space, G a subset of F with $G^s \neq \phi$ and $h : F \to \overline{R}$ a functional. Consider the following statements :

1°) For each $c < \inf h(G) = a$ there exists $\psi_c \in G^s$ satisfying (1.7) or (1.24).

2°) For each $c < a$ there exists $\psi_c \in G^s$ satisfying (1.10) or (1.35).

3°) There holds

$$\inf h(G) = \sup_{\psi \in G^s} \inf_{\substack{y \in F \\ \psi(y) \in \psi(G)}} h(y). \tag{1.37}$$

4°) For each $c < a$ there exists $\psi_c \in -G^s$ satisfying (1.7') or (1.24').

5°) For each $c < a$ there exists $\psi_c \in -G^s$ satisfying (1.10') or (1.35').

6°) There holds

$$\inf h(G) = \sup_{\psi \in -G^s} \inf_{\substack{y \in F \\ \psi(y) \in \psi(G)}} h(y). \tag{1.37'}$$

i) We have the implications 2°) $\Longrightarrow$ 1°) $\Longrightarrow$ 3°). Also, n° $\Longleftrightarrow$ $(n+3)^{o}$ (n = 1,2,3).

ii) If G and the (possibly empty) sets A_c $(c < a)$ are connected, then 1°) $\Longleftrightarrow$ 3°).

iii) If G and the (possibly empty) sets S_c $(c < a)$ are connected, then 1°) $\Longleftrightarrow$ 2°) $\Longleftrightarrow$ 3°).

The proof is similar to that of Theorem 1.1, using now Propositions 1.2 and 1.3.

Remark 1.6. a) We have always the inequalities

$$\inf h(G) \geq \sup_{\psi \in F^*} \inf_{\substack{y \in F \\ \psi(y) \in \psi(G)}} h(y) \geq$$

$$\geq \sup_{\psi \in G^s \cup \{0\}} \inf_{\substack{y \in F \\ \psi(y) \in \psi(G)}} h(y) \geq \sup_{\psi \in G^s} \inf_{\substack{y \in F \\ \psi(y) \in \psi(G)}} h(y), \tag{1.38}$$

and hence, if one of 1°), 2°) or 3°) of Theorem 1.3 holds, then

$$\inf h(G) = \sup_{\psi \in F^*} \inf_{\substack{y \in F \\ \psi(y) \in \psi(G)}} h(y) = \sup_{\psi \in G^s \cup \{0\}} \inf_{\substack{y \in F \\ \psi(y) \in \psi(G)}} h(y). \tag{1.39}$$

b) If G is connected, formula (1.37) means that

$$\inf h(G) = \sup_{\psi \in G^s} \inf h(B_{\psi,G}), \tag{1.40}$$

where $B_{\psi,G}$ is as in Remark 1.5 a) ; note that, if $\psi \in G^s$, then $B_{\psi,G}$ is proper (i.e., $B_{\psi,G} \neq F$).

c) The functionals $h : F \to \overline{R}$ such that the (possibly empty) level sets S_c $(c \in R)$ are arcwise connected, have been introduced by J. M. Ortega and W. C. Rheinboldt ([12], section 4.2), who studied uniqueness of solutions of optimization problems for such functionals. Functionals $h : F \to \overline{R}$ such that all A_c $(c \in R)$ are arcwise connected or all S_c or all A_c $(c \in R)$ are connected, and optimization problems for such functionals, have been also studied (see e.g. [1], [11], and the references therein). Clearly, quasi-convex functionals belong to each of these classes.

COROLLARY 1.5. If $G^s \neq \phi$ and G is connected, and if one of 1°),..., 6°) of Theorem 1.3 holds, then

$$\inf h(G) = \sup_{\substack{\psi \in G^s, \alpha, \beta \in \overline{R} \\ \psi(G) \subset \langle \alpha, \beta \rangle}} \inf_{\substack{y \in F \\ \psi(y) \in \langle \alpha, \beta \rangle}} h(y), \tag{1.41}$$

or, geometrically,

$$\inf h(G) = \sup_{\substack{B \in \mathcal{B} \\ B \subset G}} \inf h(B), \tag{1.42}$$

where $\mathcal{B}$ denotes the collection of all strips in F.

Proof. By (1.40) we have the inequality $\leq$ in (1.42). On the other hand, for any set $B \supset G$ we have $\inf h(G) \geq \inf h(B)$, whence (1.42) follows. □

Remark 1.7. Since every half-space is a strip, we have

$$\inf h(G) \geq \sup_{\substack{B \in \mathcal{B} \\ B \supset G}} \inf h(B) \geq \sup_{\substack{D \in \mathcal{D} \\ D \supset G}} \inf h(D), \tag{1.43}$$

and hence, by Corollary 1.2, if we have one of 1°),..., 6°) of Theorem 1.1, then (1.41), (1.42) hold (without assuming that G is connected).

From Theorems 1.2 and 1.3 there follows

THEOREM 1.4. Under the assumptions of Theorem 1.2, we have (1.37) and (1.37').

Remark 1.8. a) If G is a convex cone with vertex y_o, then, by (1.19), formula (1.37) becomes

$$\inf h(G) = \sup_{\substack{0 \neq \psi \in F^* \\ \sup \psi(G) = \psi(y_o)}} \inf_{\substack{y \in F \\ \psi(y) \in \psi(G)}} h(y), \tag{1.44}$$

and formula (1.37') changes similarly.

b) If G is a linear manifold, say $G = y_o + S$, where $y_o \in F$ and S is a linear subspace of F, then formulae (1.37) and (1.37') become

$$\inf h(G) = \sup_{0 \neq \psi \in S^{\perp}} \inf_{\substack{y\ F \\ \psi(y) = \psi(y_o)}} h(y) ; \tag{1.45}$$

geometrically, this means that

$$\inf h(G) = \sup_{\substack{H \in \mathcal{H} \\ H \supset G}} \inf h(H), \tag{1.46}$$

where $\mathcal{H}$ denotes the collection of all hyperplanes in F.

In the particular case when $G = \{y_o\}$, from Theorem 1.4 we obtain

COROLLARY 1.6. Under the assumptions of Corollary 1.4, we have

$$h(y_0) = \sup_{0\neq\psi\in F^*} \inf_{\substack{u\in F \\ \psi(y)=\psi(y_0)}} h(y). \tag{1.47}$$

I.3. RESULTS OF WEAK DUALITY IN TERMS OF CLOSED STRIPS CONTAINING G

PROPOSITION 1.4. Let F be a locally convex space, G a subset of F, $h : F \to \overline{R}$ a functional, $c \in R$ and $\psi_c \in F^*$.

i) If we have (1.7) or (1.7'), then

$$\inf_{\substack{y\in F \\ \psi_c(y)\in\overline{\psi_c(G)}}} h(y) \geq c. \tag{1.48}$$

ii) Conversely, if $\psi_c(G)$ and $\psi_c(A_c)$ are intervals in R and if (1.48) holds, then ψ_c satisfies either (1.7) or (1.7').

Proof. Part i) follows from Proposition 1.1 i), i'), observing that

$$\inf_{\substack{y\in F \\ \psi_c(y)\in\overline{\psi_c(G)}}} h(y) \geq \max\{\inf_{\substack{y\in F \\ \psi_c(y)\leq\sup\psi_c(G)}} h(y), \inf_{\substack{y\in F \\ \psi_c(y)\geq\inf\psi_c(G)}} h(y)\}. \tag{1.49}$$

Alternatively, part i) can be also obtained by replacing in the first part of the proof of Proposition 1.2 i) $\psi_c(G)$ by $\overline{\psi_c(G)}$ and observing that $\sup \overline{\psi_c(G)} = \sup \psi_c(G)$, $\inf \overline{\psi_c(G)} = \inf \psi_c(G)$.

ii) Assume that $\psi_c(G)$ and $\psi_c(A_c)$ are intervals in R and that (1.48) holds. Then, by (1.48) and $\psi_c \neq 0$,

$$A_c \subset \{y \in F \mid \psi_c(y) \notin \overline{\psi_c(G)}\}, \tag{1.50}$$

whence, since $\overline{\psi_c(G)}$ is the closed interval $[\inf \psi_c(G), \sup \psi_c(G)] \subset \overline{R}$, we obtain

$$A_c \subset \{y\in F \mid \psi_c(y) > \sup \psi_c(G)\} \cup \{y\in F \mid \psi_c(y) < \inf \psi_c(G)\}. \tag{1.51}$$

Now, by (1.48) and the obvious inequality

$$\inf_{\substack{y\in F \\ \psi_c(y)\in\psi_c(G)}} h(y) \geq \inf_{\substack{y\in F \\ \psi_c(y)\in\overline{\psi_c(G)}}} h(y) \tag{1.52}$$

we have (1.25), and hence, by Proposition 1.2 ii), A_c lies in one of the two closed half-spaces occurring in (1.27). Hence, by (1.51), A_c must lie in one of the two open half-spaces occurring in (1.51) (since the relations $y \in A_c$ and $\psi_c(y) = \sup \psi_c(G)$ or $\psi_c(y) = \inf \psi_c(G)$ would contradict (1.51)). But, this means that ψ_c satisfies either (1.7) or (1.7'), which completes the proof of Proposition 1.4.

Remark 1.9. If G is a subset of F such that $\psi(G)$ is an interval in R, then

$$\{y \in F \mid \psi(y) \in \overline{\psi(G)}\} = \overline{B}_{\psi,G} = B'_{\psi,G} , \tag{1.53}$$

where $B_{\psi,G}$ is as in Remark 1. 5 a) and where $B'_{\psi,G}$ denotes the smallest closed strip generated by ψ, containing G. Indeed, it is easy to see that

$$\psi^{-1}(\overline{\psi(G)}) = \overline{\psi^{-1}(\psi(G))} ; \tag{1.54}$$

note that (1.54) is valid for any set $G \subset F$ (and $\psi \in F^*$), even without the assumption that $\psi(G)$ is an interval in R. Now, by (1.33) and (1.54), we have the first equality in (1.53). Finally, since $\overline{\psi(G)}$ is a closed interval in $\overline{R}$, $\{y \in F \mid \psi(y) \in \overline{\psi(G)}\}$ is a closed strip, whence $B'_{\psi,G} \subset \overline{B}_{\psi,G}$; conversely, since $B'_{\psi,G}$ is a strip containing G, we have $B_{\psi,G} \subset B'_{\psi,G}$, whence $\overline{B}_{\psi,G} \subset B'_{\psi,G}$, which proves (1.53). Consequently, if $\psi_c(G)$ and $\psi_c(A_c)$ are intervals in R, then Proposition 1.4 means that we have

$$\inf_{y \in \overline{B}_{\psi_c, G}} h(y) = \inf_{y \in B'_{\psi_c,G}} h(y) \geq c \tag{1.55}$$

if and only if either ψ_c or $-\psi_c$ separates closedly G from A_c.

PROPOSITION 1.5. Let F, G, h, c and ψ_c be as in Proposition 1.4.

i) If we have (1.10) or (1.10'), then (1.48) holds.

ii) If $\psi_c(G)$ and $\psi_c(S_c)$ are intervals in R and if

$$\inf_{\substack{y \in F \\ \psi_c(y) \in \overline{\psi_c(G)}}} h(y) > c, \tag{1.56}$$

then ψ_c satisfies one of (1.10) or (1.10').

Proof. Part i) follows from $A_c \subset S_c$ and Proposition 1.4 i). The proof of part ii) is similar to that of Proposition 1.4 ii), replacing A_c by S_c and using now (1.52), (1.56) and Proposition 1.3 ii). □

THEOREM 1.5. Let F be a locally convex space, G a subset of F with $G^s \neq \phi$ and $h : F \to \overline{R}$ a functional. Consider the following statements :

1°) For each $c < \inf h(G) = a$ there exists $\psi_c \in G^s$ satisfying (1.7).

2°) For each $c < a$ there exists $\psi_c \in G^s$ satisfying (1.10).

3°) There holds

$$\inf h(G) = \sup_{\psi \in G^s} \inf_{\substack{y \in F \\ \psi(y) \in \overline{\psi(G)}}} h(y). \tag{1.57}$$

4°) For each $c < a$ there exists $\psi_c \in -G^s$ satisfying (1.7').

5°) For each $c < a$ there exists $\psi_c \in -G^s$ satisfying (1.10').

6°) There holds

$$\inf h(G) = \sup_{\psi \in -G^s} \inf_{\substack{y \in F \\ \psi(y) \in \overline{\psi(G)}}} h(y). \tag{1.57'}$$

i) We have the implications 2°) $\Longrightarrow$ 1°) $\Longrightarrow$ 3°). Also, $n° \Longleftrightarrow (n+3)°$ ($n = 1, 2, 3$).

ii) If G and the (possibly empty) sets A_c ($c < a$) are connected, then 1°) $\Longleftrightarrow$ 3°).

iii) If G and the (possibly empty) sets S_c ($c < a$) are connected, then 1°) $\Longleftrightarrow$ 2°) $\Longleftrightarrow$ 3°).

The proof is similar to that of Theorem 1.1, using now Propositions 1.4 and 1.5.

Remark 1.10. If G is connected, formula (1.57) means that in (1.40) it is enough to take the infimum of h over $\overline{B}_{\psi,G}$ (instead of $B_{\psi,G}$), or, what is the same thing, over the smallest closed strip $B'_{\psi,G}$ generated by ψ, containing G.

2. RESULTS OF STRONG DUALITY

2.0. PRELIMINARIES. Since for $c = a = \inf h(G)$ the inequalities (1.6), (1.6'), (1.25) and (1.48) become equalities (by virtue of (1.13), etc.), the main tools of obtaining results of strong duality will be Propositions 1.1, 1.2 and 1.4 with $c = a$. Thus, the role of the level sets A_c

and S_c $(c < a)$ of §1 will be played now by one single level set, namely, by

$$A_a = \{y \in F \mid h(y) < \inf h(G) = a\}. \tag{2.1}$$

Throughout the sequel, we shall denote by max a sup which is attained.

2.1. RESULTS OF STRONG DUALITY IN TERMS OF CLOSED HALF-SPACES CONTAINING G

THEOREM 2.1. Let F be a locally convex space, G a subset of F with $G^s \neq \phi$, and $h : F \to \overline{R}$ a functional. The following statements are equivalent :

1°) There exists $\psi_o \in G^s$ satisfying

$$\sup \psi_o(G) < \psi_o(y) \qquad (y \in A_a). \tag{2.2}$$

2°) There holds

$$\inf h(G) = \max_{\psi \in G^s} \inf_{\substack{y \in F \\ \psi(y) \leq \sup \psi(G)}} h(y). \tag{2.3}$$

3°) There exists $\psi_o \in -G^s$ satisfying

$$\psi_o(y) < \inf \psi_o(G) \qquad (y \in A_a). \tag{2.2'}$$

4°) There holds

$$\inf h(G) = \max_{\psi \in -G^s} \inf_{\substack{y \in F \\ \psi(y) \geq \inf \psi(G)}} h(y). \tag{2.3'}$$

Proof. 1°) $\Longrightarrow$ 2°). If $a = -\infty$ (so (2.2) is vacuously satisfied for any $\psi_o \in G^s$), then, by (1.13), we have (2.3). If $a > -\infty$, then, by 1°) and Proposition 1.1 i) (with $c = a$), we have

$$\inf_{\substack{y \in F \\ \psi_o(y) \leq \sup \psi_o(G)}} h(y) \geq a = \inf h(G),$$

whence, by (1.13), we obtain (2.3).

2°) $\Longrightarrow$ 1°). If $a = -\infty$ (so (2.3) holds by (1.13)), then (2.2) is vacuously satisfied for any $\psi_o \in G^s$. If $a > -\infty$, then, by (2.3), there exists $\psi_o \in G^s$ such that

$$a = \inf h(G) = \inf_{\substack{y \in F \\ \psi_o(y) \leq \sup \psi_o(G)}} h(y), \tag{2.4}$$

whence, by Proposition 1.1 i) (with $c = a$), we obtain (2.2).
Finally, the equivalences $n° \Longleftrightarrow (n+2)°$ ($n = 1, 2$) follow by considering $\psi'_o = -\psi_o$, $\psi' = -\psi$. This completes the proof of Theorem 2.1. □

Remark 2.1. a) Since $A_c \subset S_c \subset A_a$ for all $c < \bar{a}$, we see that (2.2) implies (1.7), (1.10) and that (2.2') implies (1.7'), (1.10').
b) Geometrically, formula (2.3) means that

$$\inf h(G) = \max_{\psi \in G^s} \inf h(D_{\psi,G}), \tag{2.5}$$

where $D_{\psi,G}$ is as in §1, Remark 1.1.

COROLLARY 2.1. If $G^s \neq \phi$ and one (and hence each) of 1°), ..., 4°) of Theorem 2.1 holds, then

$$\inf h(G) = \max_{\substack{D \in \mathcal{D} \\ D \supset G}} \inf h(D), \tag{2.6}$$

where $\mathcal{D}$ denotes the collection of all closed half-spaces in F.

The proof is similar to that of §1, Corollary 1.2, using now (2.5).
Now we shall show an important case when (2.2), respectively (2.2'), is satisfied.
We recall that the core of a set G in a linear space F is the subset of G defined by

$$\text{core } G = \{g \in G \mid \forall y \in F,\ \exists \varepsilon > 0,\ \forall \lambda \in [-\varepsilon, +\varepsilon],\ \lambda y + (1-\lambda) g \in G\}. \tag{2.7}$$

LEMMA 2.1. Let F be a locally space, G a linearly open subset of F, (i.e., G = core G), $g \in G$ and $\psi \in F^*$, $\psi \neq 0$. Then

$$\inf \psi(G) < \psi(g) < \sup \psi(G). \tag{2.8}$$

Proof. Since $g \in G$, we have the inequalities $\leq$ in (2.8). If one of them is equality, say $\psi(g) = \sup \psi(G)$, then, since $\psi \neq 0$, take $y \in F$ such that $\psi(y) < \sup \psi(G)$, and let

$$g_\lambda = \frac{1}{1-\lambda} g - \frac{\lambda}{1-\lambda} y \qquad (0 < \lambda < 1). \tag{2.9}$$

Then, since $g \in G$ = core G, for λ sufficiently small we have $g_\lambda \in G$.

But, from $\psi \in F^*$, $\psi(g) = \sup \psi(G)$ and $\psi(y) < \sup \psi(G)$, we get

$$\psi(g_\lambda) = \frac{1}{1-\lambda} \psi(g) - \frac{\lambda}{1-\lambda} \psi(y) > \sup \psi(G),$$

a contradiction. This proves the second inequality in (2.8), for any $\psi \in F^*$, $\psi \neq 0$, whence also the first one (which can be written in the form $\sup (-\psi)(G) > (-\psi)(g)$). □

LEMMA 2.2. Let F be a locally convex space, G a subset of F, $h : F \to \overline{R}$ a functional such that A_a is linearly open, and $\psi_o \in F^*$, $\psi_o \neq 0$.

i) If

$$\sup \psi_o(G) \leq \inf \psi_o(A_a), \tag{2.10}$$

then ψ_o satisfies (2.2).

i') If

$$\sup \psi_o(A_a) \leq \inf \psi_o(G), \tag{2.10'}$$

then ψ_o satisfies (2.2').

Proof. i) Since A_a is linearly open, by (2.10) and Lemma 2.1 we have

$$\sup \psi_o(G) \leq \inf \psi_o(A_a) < \psi_o(y) \qquad (y \in A_a), \tag{2.11}$$

so (2.2) holds, which proves i). The proof of i') is similar. □

Remark 2.2. In particular, if h is upper semi-continuous along lines, i.e., if for each line

$$L = L_{y_1,y_2} = \{\lambda y_2 + (1-\lambda)y_1 \in F \mid -\infty < \lambda < +\infty\}, \tag{2.12}$$

where y_1, $y_2 \in F$, the restriction $h|_L$ is upper semi-continuous, which is an assumption usually made in the theory of quasi-convex optimization (see e.g. [10]), then A_a is linearly open. Indeed, if $y_1 \in A_a$ and $y_2 \in F$, then, by our assumption, the set

$$L \cap A_a = \{\lambda y_2 + (1-\lambda)y_1 \in L \mid h(\lambda y_2 + (1-\lambda)y_1) < a\} \tag{2.13}$$

is an open interval, containing y_1 (take $\lambda = 0$). Consequently $y_1 \in \text{core } A_a$, so A_a is linearly open, which proves our assertion. Moreover, there holds also the following converse type result : if all level sets $A_r = \{y \in F \mid h(y) < r\}$ $(r \in R)$ are linearly open, then h is upper semicontinuous along lines. Indeed, if there exists $r \in R$ such that the interval $L \cap A_r$ contains one of its endpoints, say y_o, then $y_o \in A_r \setminus \text{core } A_r$, so A_r is not linearly open.

It will be worth while to state separately the following sufficient condition in order to have (2.3), (2.3'), (2.6) :

THEOREM 2.2. Let F be a locally convex space, G a convex subset of F with $G^s \neq \phi$ and $h : F \to \overline{R}$ a functional, with A_a convex and open. Then we have 1°) of Theorem 2.1, whence also (2.3), (2.3'), (2.6).

Proof. If $A_a = \phi$, then, by $G^s \neq \phi$, we have 1°) of Theorem 1.1 (for any $\psi_o \in G^s$). On the other hand, if $A_a \neq \phi$, then G and A_a are non-empty convex subsets of F, with A_a open and, clearly, $G \cap A_a = \phi$. Hence, there exists (see e.g. [3], p. 417) a functional $\psi_o \in F^*$, $\psi_o \neq 0$, satisfying (2.10). Then, by Lemma 2.2, we have (2.2), whence $\psi_o \in G^s$. Thus, we have 1°) of Theorem 2.1, whence also (2.3), (2.3'), (2.6), which completes the proof of Theorem 2.2. □

Remark 2.3. a) In particular, the assumption on h, of Theorem 2.2, is satisfied if $h : F \to \overline{R}$ is convex and upper semi-continuous. For $h : F \to \overline{R}$ convex and upper semi-continuous at some $g_o \in G$ with $h(g_o) < +\infty$, the part (2.6) of Theorem 2.2 has been obtained by P.-J. Laurent and B. Martinet ([9], p. 167, formula (13) ; see also [8], p. 430, Lemma 1). This result, with the conclusions (2.3), (2.3'), (2.6), can be also deduced from Theorem 2.2.

b) Under the above assumptions (of Remark 2.3 a)) we have, by a consequence of the Fenchel-Rockafellar duality theorem (see [16], Theorem 2.1 and the end of §1),

$$\inf h(G) = \max_{\psi \in F^*} \inf_{y \in F} \{h(y)+\psi(y)-\sup \psi(G)\}, \tag{2.14}$$

which, by the obvious inequalities

$$\inf h(G) \geq \sup_{\psi \in G^s} \inf_{\substack{y \in F \\ \psi(y) \leq \sup\psi(G)}} h(y) \geq \max_{\psi \in F^*} \inf_{y \in F} \{h(y)+\psi(y)-\sup\psi(G)\}, \tag{2.15}$$

implies again (2.3), whence also (2.3'), (2.6). For this method of obtaining duality theorems from Lagrangian ones, see [17], the part added in proof and [18], §4.

In view of its importance for applications, let us state now separately the particular case $G = \{y_o\}$ of Corollary 2.1 and Remark 2.3 a) :

COROLLARY 2.2. Let F be a locally convex space, $y_o \in F$, and $h : F \to \overline{R}$ a functional, such that the level set $A_{h(y_o)}$ is convex and open ; or, let $h : F \to \overline{R}$ be a convex functional, which is upper semi-continuous at y_o and such that $h(y_o) < +\infty$. Then

$$h(y_o) = \max_{0 \neq \psi \in F^*} \inf_{\substack{y \in F \\ \psi(y) \geq \psi(y_o)}} h(y). \tag{2.16}$$

2.2. RESULTS OF STRONG DUALITY IN TERMS OF STRIPS CONTAINING G

THEOREM 2.3. Let F be a locally convex space, G a subset of F with $G^s \neq \phi$, and $h : F \to \overline{R}$ a functional. Consider the following statements :

1°) There exists $\psi_o \in G^s$ satisfying either (2.2) or

$$\psi_o(g) < \inf \psi_o(A_a) \quad (g \in G). \tag{2.17}$$

2°) There holds

$$\inf h(G) = \max_{\psi \in G^s} \inf_{\substack{y \in F \\ \psi(y) \in \psi(G)}} h(y). \tag{2.18}$$

3°) There exists $\psi_o \in -G^s$ satisfying either (2.2') or

$$\sup \psi_o(A_a) < \psi_o(g) \quad (g \in G). \tag{2.17'}$$

4°) There holds

$$\inf h(G) = \max_{\psi_o \in -G^s} \inf_{\substack{y \in F \\ \psi(y) \in \psi(G)}} h(y). \tag{2.18'}$$

i) We have the implication 1°) $\Longrightarrow$ 2°). Also, n° $\Longleftrightarrow$ (n+2)° (n = 1, 2).

ii) If G and the (possibly empty) set A_a are connected, then 1°) $\Longleftrightarrow$ 2°).

The proof is similar to that of Theorem 2.1, using now Proposition 1.2 (with c = a).

Similarly to Lemma 2.2, we shall show now an important case in which (2.17), respectively (2.17'), is satisfied.

LEMMA 2.3. Let F be a locally convex space, G a linearly open subset of F, $h : F \to \overline{R}$ a functional and $\psi_o \in F^*$, $\psi_o \neq 0$.

i) If we have (2.10), then ψ_o satisfies (2.17).

i') If we have (2.10'), then ψ_o satisfies (2.17').

Proof. i) Since G is linearly open, by Lemma 2.1 and (2.10) we have

$$\psi_o(g) < \sup \psi_o(G) \leq \inf \psi_o(A_a) \qquad (g \in G),$$

so (2.17) holds, which proves i). The proof of i') is similar. □

Similarly to the proof of Theorem 2.2, one obtains

THEOREM 2.4. Let F be a locally convex space , G a convex subset of F with $G^s \neq \phi$ and $h : F \to \overline{R}$ a functional, whith A_a convex, such that either G or A_a is open. Then we have 1°) of Theorem 2.3, whence also (2.18) and (2.18').

In the particular case when $G = \{y_o\}$, from Theorem 2.4 and Remark 2.3 a) we obtain

COROLLARY 2.3. Under the assumptions of Corollary 2.2, we have

$$h(y_o) = \max_{0 \neq \psi \in F^*} \inf_{\substack{y \in F \\ \psi(y)=\psi(y_o)}} h(y). \tag{2.19}$$

2.3. RESULTS OF STRONG DUALITY IN TERMS OF CLOSED STRIPS CONTAINING G

THEOREM 2.5. Let F be a locally convex space, G a subset of F with $G^s \neq \phi$ and $h : F \to \overline{R}$ a functional. Consider the following statements :

1°) There exists $\psi_o \in G^s$ satisfying (2.2).

2°) There holds

$$\inf h(G) = \max_{\psi \in G^s} \inf_{\substack{y \in F \\ \psi(y) \in \overline{\psi(G)}}} h(y). \tag{2.20}$$

3°) There exists $\psi_o \in -G^s$ satisfying (2.2').

4°) There holds

$$\inf h(G) = \max_{\psi \in -G^S} \inf_{\substack{y \in F \\ \psi(y) \in \overline{\psi(G)}}} h(y). \qquad (2.20')$$

i) We have the implication 1°) $\Longrightarrow$ 2°). Also, n° $\Longleftrightarrow$ (n+2)° (n=1,2)

ii) If G and the (possibly empty) set A_a are connected, then 1°) $\Longleftrightarrow$ 2°).

The proof is similar to that of Theorem 2.1, using now Proposition 1.4 (with c = a).

Let us give now an important case in which (2.17) implies (2.2) and hence (2.20).

THEOREM 2.6. Let F be a locally convex space, G a subset of F, $h : F \to \overline{R}$ a convex functional, with $a = \inf h(G) < +\infty$, and $\psi_o \in F^*$. If (2.17) holds, then we have (2.2) and hence also (2.20).

Proof. If (2.2) does not hold, then there exists $y_o \in A_a$ such that

$$\psi_o(y_o) \leq \sup \psi_o(G). \qquad (2.21)$$

Let $\varepsilon > 0$ be such that

$$h(y_o) + \varepsilon < a \qquad (2.22)$$

and choose $g_\varepsilon \in G$ such that

$$h(g_\varepsilon) \leq \inf h(G) + \varepsilon = a + \varepsilon. \qquad (2.23)$$

Then, since h is convex and since $a < +\infty$, for any λ with $\frac{1}{2} \leq \lambda \leq 1$ we have

$$h(\lambda y_o+(1-\lambda)g_\varepsilon) \leq \lambda h(y_o)+(1-\lambda)h(g_\varepsilon) < \lambda(a-\varepsilon)+(1-\lambda)(a+\varepsilon) =$$
$$= (1-2\lambda)\,\varepsilon + a \leq a,$$

so $\lambda y_o + (1-\lambda)\, g_\varepsilon \in A_a$. But, by (2.21) and (2.17) we have

$$\psi_o(y_o) \leq \sup \psi_o(G) \leq \inf \psi_o(A_a),$$

$$\psi_o(g_\varepsilon) < \inf \psi_o(A_a),$$

whence

$$\psi_o(\lambda y_o+(1-\lambda)g_\varepsilon) = \lambda\psi_o(y_o)+(1-\lambda)\psi_o(g_\varepsilon) < \inf \psi_o(A_a),$$

in contradiction with $\lambda y_o + (1-\lambda)\, g_\varepsilon \in A_a$. This proves (2.2). Hence, by Theorem 2.5, we have (2.20), which completes the proof of Theorem 2.6. □

Remark 2.4. a) One cannot omit in Theorem 2.6 the assumption that $a < +\infty$, as shown by the following simple example : Let $F = R^2$ (the Euclidean plane), let

$$G = \{(\eta_1, \eta_2) \in F \mid \max(|\eta_1 + 1|, |\eta_2|) < 1\}, \tag{2.24}$$

and let $h = \chi_M$, the indicator functional of the convex set

$$M = \{(\eta_1, \eta_2) \in F \mid \max(|\eta_1 - 1|, |\eta_2|) \leq 1\}. \tag{2.25}$$

Then $a = \inf h(G) = +\infty$ (since $G \cap M = \phi$), whence $A_a = M$, and for $\psi_o \in F^*$ defined by

$$\psi_o(y) = \eta_1 \qquad (y = (\eta_1, \eta_2) \in F) \tag{2.26}$$

we have (2.17), but neither (2.2), nor (2.20).

b) One cannot replace in Theorem 2.6 the assumption of convexity of h by the weaker assumption that h is quasi-convex, as can be seen by taking in a) above, instead of $h = \chi_M$, the characteristic functional of M, that is, the quasi-convex functional

$$h(y) = \begin{cases} 0 & \text{if } y \in M \\ 1 & \text{if } y \in F \setminus M. \end{cases} \tag{2.27}$$

2.4. COMPLEMENTARY REMARKS

Remark 2.5. Propositions 1.1, 1.2 and 1.4, with c = a, may be regarded as characterizations of "optimal functionals", i.e., of the functionals ψ_o for which the max in the right hand side of one of our formulae of strong duality (2.3), (2.3'), (2.18), (2.18'), (2.20) or (2.20') is attained. We shall give elsewhere some further characterizations of optimal functionals.

Remark 2.6. The optimal functionals ψ_o (see Remark 2.5) are nothing else than the "solutions" of the problems of maximizing one of the functionals

$$\mu_1(\psi) = \inf_{\substack{y \in F \\ \psi(y) \leq \sup \psi(G)}} h(y) \qquad (\psi \in F^*), \tag{2.28}$$

$$\mu_1'(\psi) = \inf_{\substack{y\in F \\ \psi(y)\geq \inf \psi(G)}} h(y)=\mu_1(-\psi) \quad (\psi \in F^*), \tag{2.28'}$$

$$\mu_2(\psi) = \inf_{\substack{y\in F \\ \psi(y) \in \psi(G)}} h(y) \qquad (\psi \in F^*), \tag{2.29}$$

$$\mu_3(\psi) = \inf_{\substack{y\in F \\ \psi(y) \in \overline{\psi(G)}}} h(y) \qquad (\psi \in F^*), \tag{2.30}$$

over one of the sets G^s or $-G^s$ respectively, which may be called "dual problems" to the optimization problem (0.1). Thus, the above mentioned results on optimal functionals can be also expressed in terms of solutions of various dual problems.

Finally, let us make some observations on the separation of G and A_a by ψ_o or $-\psi_o$, occurring in our results of strong duality.

Remark 2.7. If

$$\overline{G} \cap \overline{A}_a \neq \phi, \tag{2.31}$$

then the inequalities (2.10) and (2.10') become equalities, that is, respectively,

$$\sup \psi_o(G) = \inf \psi_o(A_a), \tag{2.32}$$

$$\sup \psi_o(A_a) = \inf \psi_o(G). \tag{2.32'}$$

Indeed, if $y_o \in \overline{G} \cap \overline{A}_a$, then since $\psi_o \in F^*$, from (2.10) we obtain

$$\psi_o(y_o) \leq \sup \psi_o(\overline{G}) = \sup \psi_o(G) \leq \inf \psi_o(A_a) = \\ = \inf \psi_o(\overline{A}_a) \leq \psi_o(y_o),$$

whence (2.32), and, similarly, (2.10') implies (2.32'). One can give some simple sufficient conditions for (2.31), ensuring thus (2.32) or (2.32') in the above results involving one of (2.10) or (2.10'). Let us first observe

LEMMA 2.4. Let F be a locally convex space, $c \in R$ and $h : F \to \overline{R}$ a convex functional, with $A_c \neq \phi$. Then

$$S_c \subset \overline{A}_c. \tag{2.33}$$

Proof. Let $y_o \in S_c$ and, since $A_c \neq \phi$, let $y' \in A_c$. Put

$$y_n = \frac{1}{n} y' + \frac{n-1}{n} y_o \quad (n=1,2,\ldots). \tag{2.34}$$

Then $y_n \to y_o$ as $n \to \infty$ and, since h is convex and $c < +\infty$, from $y' \in A_c$ and $y_o \in S_c$ we obtain

$$h(y_n) \le \frac{1}{n} h(y') + \frac{n-1}{n} h(y_o) < c \quad (n=1,2,\ldots), \tag{2.35}$$

that is, $y_n \in A_c$ $(n=1,2,\ldots)$. Thus, $y_o \in \overline{A}_c$, which proves (2.33). □

Remark 2.8. Since $A_c \subset S_c$, we have $\overline{A}_c \subset \overline{S}_c$ and hence, if h is lower semi-continuous, then, under the assumptions of Lemma 2.4 we have

$$S_c = \overline{A}_c. \tag{2.36}$$

PROPOSITION 2.1. Let F be a locally convex space, G a subset of F and $h : F \to \overline{R}$ a convex functional, with $a = \inf h(G) < +\infty$ and $A_a \neq \phi$. If there exists $y_o \in \overline{G}$ such that

$$h(y_o) \le \inf h(G) = a, \tag{2.37}$$

then (2.31) holds. Hence, in this case, if $\psi_o \in F^*$ satisfies (2.10) or (2.10') (in particular, if ψ_o satisfies one of (2.2), (2.17), (2.2'), (2.17')), then we have either (2.32) or (2.32') respectively.

Proof. By Lemma 2.4, for $y_o \in \overline{G}$ satisfying (2.37) we have

$$y_o \in \overline{G} \cap S_a \subset \overline{G} \cap \overline{A}_a, \tag{2.38}$$

so (2.31) holds. Hence, by Remark 2.7, the conclusion follows. □

Remark 2.9. a) In general, there need not exist $y_o \in \overline{G}$ satisfying (2.37), as shown e.g. by a closed subset G of a normed linear space F and $h(y') = \|y' - y\|$ $(y \in F)$, where $y' \in F$ has no best approximation in G. However, if either i) every sequence $\{g_n\} \subset G$ with $h(g_n) \to \inf h(G) = a < +\infty$ contains a subsequence $\{g_{n_k}\}$ converging to some $y_o \in F$ and h is lower semi-continuous, or ii) $G = \{y_o\}$, a singleton, then there exists $y_o \in \overline{G}$ satisfying (2.37). Indeed, since $a < +\infty$, for each n there exists $g_n \in G$ such that $a \le h(g_n) \le a + \frac{1}{n}$, whence $h(g_n) \to a$. Then, in case i), $\{g_n\}$ contains a subsequence $\{g_{n_k}\}$ converging to some $y_o \in F$, whence, since h is lower semi-continuous, $h(y_o) \leq \underline{\lim}\, h(g_{n_k}) = a$, so (2.37) holds. In case ii), the statement is obviously true.

b) One cannot omit in Proposition 2.1 the assumption that $a = \inf h(G) < +\infty$, as shown by $h = \chi_M$, the indicator functional of any convex set M with $\overline{M} \neq F$ and by $G = \{g_o\}$ (a singleton), where $g_o \in F \setminus M$.

c) One cannot replace in Proposition 2.1 the assumption of convexity of h by the weaker assumption that h is quasi-convex, as shown by taking, instead of the functional $h = \chi_M$ in b) above, the quasi-convex functional (2.27).

Remark 2.10. a) Let us say that a subset G of F is F^*-connected, if $\psi(G) \subset R$ is an interval $\langle\alpha, \beta\rangle$, for each $\psi \in F^*$. Then every connected set is F^*-connected, but the converse is not true, as shown by simple examples. One can replace everywhere in the above the word "connected" by : F^*-connected.

b) One can replace in all results of this paper the set G^s by the smaller set $G^o \setminus \{0\}$. Indeed, if $1 < \sup \psi(G) < +\infty$, then for $\psi' = \frac{1}{\sup \psi(G)} \psi$ we have $\psi' \in G^o \setminus \{0\}$ and $\psi'(y) \leq \sup \psi'(G) = 1 \Longleftrightarrow \psi(y) \leq \sup \psi(G)$, $\psi'(y) \in \psi'(G) \Longleftrightarrow \psi(y) \in \psi(G)$, $\psi'(y) \in \overline{\psi'(G)} \Longleftrightarrow \psi(y) \in \overline{\psi(G)}$.

c) If $\varepsilon > 0$, one can add in Theorem 1.1 the following equivalent condition :

7°) For each $c \in (a-\varepsilon, a)$ there exists $\psi_c \in G^s$ satisfying (1.7). Indeed, the implication 1°) $\Longrightarrow$ 7°) is obvious, while the implication 7°) $\Longrightarrow$ 1°) follows by observing that if ψ_c satisfies (1.7) and if $c' < c$, then one can take $\psi_{c'} = \psi_c$ (since $A_{c'} \subset A_c$). Similar remarks can be also made for (1.10), (1.24), (1.35) and (1.7'), etc, in the theorems of §1.

REFERENCES

1. M. Avriel and I. Zang, Generalized arcwise connected functions and characterizations of local-global minimum properties. J. Optimiz. Theory Appl. 32 (1980), 407-425.

2. J.-P. Crouzeix, Contributions à l'étude des fonctions quasi-convexes. Thèse, Université de Clermont-Ferrand (France), Octobre 1977.

3. N. Dunford and J. Schwartz, Linear operators. Part I : General theory. Pure and Appl. Math. 7, Interscience, New York-London, 1958.

4. F. Glover, A multiphase-dual algorithm for the zero-one integer programming problem. Opns. Res. 13 (1965), 879-919.

5. H. J. Greenberg and W. P. Pierskalla, Surrogate mathematical programming. Opns. Res. 18 (1970), 924-939.

6. H. J. Greenberg and W. P. Pierskalla, Quasi-conjugate functions and surrogate duality. Cahiers Centre d'Et. Rech. Opér. 15 (1973), 437-448.

7. V. Klee, Separation and support properties of convex sets-a survey. In : Control theory and the calculus of variations, 235-303. Academic Press, New York, 1969.

8. P.-J. Laurent, Construction of spline functions in a convex set. In : Approximation with special emphasis on spline functions, 415-446. Academic Press, New York, 1969.

9. P.-J. Laurent et B. Martinet, Méthodes duales pour le calcul du minimum d'une fonction convexe sur une intersection de convexes. In : Symposium on optimization, Nice, 1969. Lecture Notes in Math. 132, Springer, 1970.

10. D. G. Luenberger, Quasi-convex programming. SIAM J. Appl. Math. 16 (1968), 1090-1095.

11. D. H. Martin, Connected level sets and uniqueness in optimization. Part 1, Nat. Res. Inst. Math. Sci. Pretoria, CSIR Spec. Rep. WISK-276, 1977 ; Part 2, ibid. WISK-277, 1977.

12. J. M. Ortega and W. C. Rheinboldt, Iterative solution of nonlinear equations in several variables. Acad. Press, New York, 1970.

13. R. T. Rockafellar, Convex analysis. Princeton Univ. Press, Princeton, New-Jersey, 1970.

14. G. Š. Rubinštein, Dual extremal problems. Doklady Akad. Nauk SSSR 152 (1963), 288-291 (Russian).

15. G. Š. Rubinštein, Duality in mathematical programming and some problems of convex analysis. Uspehi Mat. Nauk 25, 5 (155) (1970), 171-201 (Russian).

16. I. Singer, Some new applications of the Fenchel-Rockafellar duality theorem : Lagrange multiplier theorems and hyperplane theorems for convex optimization and best approximation. Nonlinear Anal. Theory, Methods Appl. 3 (1979), 239-248.

17. I. Singer, Duality theorems for linear systems and convex systems. J. Math. Anal. Appl. 76 (1980), 339-368.

18. I. Singer, Duality theorems for perturbed convex optimization. J. Math. Anal. Appl. 81 (1981), 437-452.

A GENERALIZED CONCEPT OF CONJUGATION

J.-E. Martinez-Legaz

University of Barcelona
Barcelona, Spain

In convex analysis, the conjugate function of $f : R^n \to \overline{R}$ is defined by $f^*(x^*) = \sup\{<x^*, x> - f(x)\}$. If H respresents the family of functions $\{h_b \mid b \in \overline{R}\}$, with $h_b(t) = t - b$, $f^*(x^*)$ is characterized by the relation :

$$h_{f^*(x^*)} = \sup\{h_b \mid h_b \in H,\ h_b(<x^*, x>) \le f(x) \quad \forall x\}.$$

This suggests to extend the notion of conjugation by considering other families in such a way that the essential properties of the classical theory subsist. In particular, taking the family of continuous from the left non-decreasing functions one obtains results equivalent in a sense to the theories for quasiconvex conjugation given by Crouzeix and by Greenberg and Pierskalla respectively [3]. In the same way that a closed convex function is a supremum of affine functions, quasiaffine functions are defined being in a similar relation with the quasiconvex ones.

This general notion of conjugation is also used to construct dual problems in mathematical programming and Lagrangians from which dual problems can be derived.

I. H-CONJUGATES

Let H be a family of extended real valued functions of a single variable $h : R \to \overline{R}$, closed under pointwise supremum, that is, such that $H' \subset H$ implies $\sup\{h \mid h \in H'\} \in H$. A function $f : R^n \to \overline{R}$ will be called H-convex if it can be expressed as the supremum of those obtained composing some $h \in H$ with some linear function, i.e., if one has for all x_o :

$$f(x_o) = \sup\{h(\langle x^*, x_o\rangle) \mid h \in H,\ x^* \in R^n,\ h(\langle x^*, x\rangle) \le f(x) \quad \forall x\}.$$

If $\Phi = \{\varphi : R^n \to \overline{R} \mid \varphi(x) = h(\langle x^*, x\rangle)$ for some $h \in H,\ x^* \in R^n\}$, H-convex functions coincide with Φ-convex functions in the sense of Dolecki and Kurcyusz [5]. The supremum of H-convex functions is again an H-convex function ; then for each $f : R^n \to \overline{R}$ there exists $f^o : R^n \to \overline{R}$, the greatest H-convex function majorized by f. f^o will be called the H-convex hull of f.

DEFINITION. For $f : R^n \to \overline{R}$, its H-conjugate is $f^c : R^n \to H$ given by $f^c(x^*) = \sup\{h \mid h \in H,\ h(\langle x^*, x\rangle) \le f(x)$ for all $x \in R^n\}$. Inversely, given $g : R^n \to H$, its conjugate function $g^\nabla : R^n \to \overline{R}$ is defined by $g^\nabla(x) = \sup_{x^*} g(x^*)(\langle x^*, x\rangle)$.

As an immediate consequence of these definitions we have the following properties :

PROPOSITION 1. a) $f^c(x^*)(\langle x^*, x\rangle) \le f(x)$ for all $x^*, x \in R^n$;
b) $f^c = f^{oc}$; c) g^∇ is H-convex.

PROPOSITION 2. The H-convex hull of any function f is equal to the second conjugate $f^{c\nabla}$. In other terms $f^{c\nabla} = f^o$.

Proof. $f^c(x^*) = \sup\{h \mid h \in H,\ h(\langle x^*, x\rangle) \le f(x) \quad \forall x\}$. Applying this equality to $\langle x^*, x_o\rangle$ we obtain

$$f^c(x^*)(\langle x^*, x_o\rangle) = \sup\{h(\langle x^*, x_o\rangle) \mid h \in H,\ h(\langle x^*, x\rangle) \le f(x) \quad \forall x\}.$$

Finally, $f^{c\nabla}(x_o) = \sup_{x^*} f^c(x^*)(\langle x^*, x_o\rangle) =$

$$= \sup_{x^*, h}\{h(\langle x\ , x_o\rangle) \mid x^* \in R^n,\ h \in H,\ h(\langle x^*, x\rangle) \le f(x)\ \forall x\} = f^o(x_o). \ \square$$

COROLLARY 3. f is H-convex if and only if $f = f^{c\nabla}$.

DEFINITION. x^* is a H-subgradient of f at $x_o \in R^n$ if there is some $h_o \in H$ such that $h_o(<x^*, x_o>) = f(x_o)$, $h_o(<x^*, x>) \leq f(x)$ for all x.

The set of all H-subgradients of f at x_o will be denoted by $\partial_H f(x_o)$.

PROPOSITION 4. x^* is an H-subgradient of f at x_o if and only if $f^c(x^*)(<x^*, x_o>) = f(x_o)$.

Proof. $f^c(x^*)(<x^*, x_o>) = f(x_o)$, by Proposition 1.a) we can take $h_o = f^c(x^*)$ in the definition before. Conversely, if x^* is an H-subgradient of f at x_o we have :

$$f(x_o) = h_o(<x^*,x_o>) \leq \sup\{h(<x^*,x_o>) \mid h \in H,\ h(<x^*,x>) \leq f(x) \quad \forall x\}$$

$$= f^c(x^*)(<x^*, x_o>) \leq f(x_o). \quad \square$$

PROPOSITION 5. Let us assume that H verifies the following property : $\forall \alpha \in R \quad \exists h_\alpha \in H \mid h_\alpha(0) = \alpha$. Then, $0 \in \partial_H f(x_o) \iff f(x_o) = \inf_x f(x)$.

Proof. $f^c(0)(0) = \sup\{h(0) \mid h \in H,\ h(0) \leq f(x)\ \forall x\} \leq f(x)$ for all x. Hence $f^c(0)(0) = \inf_x f(x)$. By the assumption made on H, we have for $\alpha = \inf_x f(x)$,

$$\inf_x f(x) = h_\alpha(0) = \sup\{h(0) \mid h \in H,\ h(0) \leq f(x) \quad \forall x\} = f^c(0)(0).$$

Then, we have proved that $f^c(0)(0) = \inf_x f(x)$, and the statement is obtained applying Proposition 4. $\square$

PROPOSITION 6. If $\partial_H f(x_o) \neq \phi$, then f is H-convex at x_o. If f is H-convex at x_o, then $\partial_H f(x_o) = \partial_H f^o(x_o)$.

Proof. a) Taking $x^* \in \partial_H f(x_o)$, by Propositions 3 and 1 :

$$f(x_o) = f^c(x^*)(<x^*, x_o>) = f^{oc}(x^*)(<x^*, x_o>) \leq f^o(x_o) \leq f(x_o).$$

b) $x^* \in \partial_H f(x_o) \iff f^c(x^*)(<x^*,x_o>) = f(x_o) \iff f^{oc}(x^*)(<x^*,x_o>) = f^c(x_o) \iff x^* \in \partial_H f^o(x_o)$. $\square$

The converse of part a) of Proposition 6 is not true. A counter-example will be given at the end of section 5.

PROPOSITION 7. $H_1 \subset H_2 \Longrightarrow \partial_{H_1} f(x_o) \subset \partial_{H_2} f(x_o)$.

Proof. It follows directly from the definition of H-subgradient. □

II. H-DUALITY

Let us consider the family of mathematical programming problems "$\underset{x \in C}{\text{minimize}}\ \phi(x, w)$" depending on $w \in R^m$, obtained by perturbing the original primal problem "$\underset{x \in C}{\text{minimize}}\ \phi(x, 0)$". By Proposition 1.a) we have :

$$\phi^c(x^*, w^*)(\langle x^*, x\rangle + \langle w^*, w\rangle) \leq \phi(x, w) \text{ for all } x, w, x^*, w^*.$$

In particular, setting $x^* = 0$, $w = 0$ we obtain :

$$\phi^c(0, w^*)(0) \leq \phi(x, 0) \text{ for all } x, w^*.$$

This motivates the following definition :

DEFINITION. The problem "$\underset{w^* \in R^m}{\text{maximize}}\ \phi^c(0, w^*)(0)$" will be called the dual problem.

The last inequality can be restated as the following theorem :

THEOREM 8 (Weak Duality). $\inf\limits_x \phi(x, 0) \leq \sup\limits_{w^*} \phi^c(0, w^*)(0)$.

COROLLARY 9. If $\phi(x_o, 0) = \phi^c(0, w_o^*)(0)$, then x_o and w_o^* are optimal solutions of the primal and dual problems respectively.

DEFINITION. The perturbation function is defined by $P(w) = \inf\limits_x \phi(x, w)$.

PROPOSITION 10. $\phi^c(0, w^*) = P^c(w^*)$.

Proof.
$$\begin{aligned}\phi^c(0, w^*) &= \sup\{h \mid h \in H,\ h(\langle w^*, w\rangle) \leq \phi(x, w)\ \forall x, w\} = \\ &= \sup\{h \mid h \in H,\ h(\langle w^*, w\rangle) \leq \inf_x \phi(x, w)\ \forall x\} = \\ &= \sup\{h \mid h \in H,\ h(\langle w^*, w\rangle) \leq P(w)\ \forall w\} = P^c(w^*).\end{aligned}$$
□

According with this proposition, the dual problem can be written "maximize $P^c(w^*)(0)$".

THEOREM 11. $\inf_x \phi(x, 0) = \sup_{w^*} \phi^c(0, w^*)(0)$ if and only if P is H-convex at 0.

Proof. By définition of P, $\inf_x \phi(x, 0) = P(0)$. By Proposition 9, $\sup \phi^c(0, w^*)(0) = \sup_{w^*} P^c(w^*)(0) = P^{c\nabla}(0)$. The statement is obtained applying Proposition 2. □

PROPOSITION 12. The set of optimal solutions of the dual problem is $\partial_H P^o(0)$.

Proof. w_o^* is an optimal solution of the dual problem means that $P^c(w_o^*)(0) = \sup_{w^*} P^c(w^*)(0) = P^{c\nabla}(0) = P^o(0)$. Since $P^c = P^{oc}$, the last equality is the equivalent definition of H-subgradient for w_o^* given in Proposition 4. □

THEOREM 13 (Strong Duality). The dual problem has an optimal solution and the optimal values of the primal and dual problems are equal if and only if $\partial_H P(0) \neq \phi$. In this case, the set of optimal dual solutions is $\partial_H P(0)$.

Proof. If the optimal values of both problems are equal, by Theorem 11 P is H-convex at 0 and by Proposition 6.b) $\partial_H P(0) = \partial_H P^o(0)$. Since there is an optimal solution, by Proposition 12 this set is not empty. Now, let us suppose that $\partial_H P(0) \neq \phi$. By Proposition 6, $\partial_H P(0) = \partial_H P^o(0)$ and P is H-convex at 0. Then we can apply Proposition 12 and Theorem 11. □

III. H-LAGRANGIANS

Let us call ϕ_x the function from R^m into $\overline{R}$ defined by $\phi_x(w) = \phi(x, w)$.

DEFINITION. The H-lagrangian associated with the family of perturbed problems is $L : C \times R^m \to \overline{R}$ given by $L(x, w^*) = \phi_x^c(w^*)(0)$.

PROPOSITION 14. $\phi(x, 0) \geq \sup_{w^*} L(x, w^*)$ for each $x \in R^n$. The equality holds if and only if ϕ_x is H-convex at 0.

Proof. $\phi(x,0) = \phi_x(0)$ and $\sup_{w^*} L(x,w^*) = \sup_{w^*} \phi_x^c(w^*)(0) = \phi_x^{c\nabla}(0) =$

$= \phi_x^o(0)$. □

COROLLARY 15. $P(0) \geq \inf_x \sup_{w^*} L(x, w^*)$. If ϕ_x is H-convex at 0 $\forall x$, the equality holds.

PROPOSITION 16. $\phi^c(0, w^*)(0) \leq \inf_x L(x, w^*)$ for each $w^* \in R^m$. If H is closed under pointwise infimum, the equality holds.

Proof. $\phi^c(0, w^*)$ belongs to H and by Proposition 1.a) $\phi^c(0, w^*)(\langle w^*, w\rangle) \leq \phi(x, w)$ $\forall x, w$. On the other hand, given any $x \in R^n$, we have :

$$\phi_x^c(w^*) = \sup\{h \mid h \in H,\ h(\langle w^*, w\rangle) \leq \phi_x(0) \ \ \forall w\} =$$

$$= \sup\{h \mid h \in H,\ h(\langle w^*, w\rangle) \leq \phi(x, w) \ \ \forall w\}.$$

Then $\phi^c(0, w^*) \leq \phi_x^c(w^*)$ for all x, that is, $\phi^c(0, w^*) \leq \inf_x \phi_x^c(w^*)$, and the first part of the statement is obtained applying this inequality at 0. Now, let us suppose that H is closed under pointwise infimum. We only need prove the converse of last inequality. The assumption on H implies that $\inf_x \phi_x^c(w^*)$ belongs to H. Moreover

$$\inf_x \phi_x^c(w^*)(\langle w^*, w\rangle) \leq \inf_x \phi_x(w) = \inf_x \phi(x, w) \leq \phi(x, w) \text{ for}$$

all x, w. Hence $\inf_x \phi_x^c(w^*) \leq \phi^c(0, w^*)$. □

COROLLARY 17. $P^o(0) \leq \sup_{w^*} \inf_x L(x, w^*)$. If H is closed under pointwise infimum the equality holds.

Proof. $P^o(0) = P^{c\nabla}(0) = \sup_{w^*} P^c(w^*)(0) = \sup_w \phi^c(0, w^*)(0)$. □

PROPOSITION 18. P H-convex at 0 implies that $\sup_{w^*} \inf_x L(x,w^*) = P(0)$. If h is closed under pointwise infimum, the conserve is also true.

Proof. If P is H-convex at 0, by Corollaries 17 and 14,

$$P(0) = P^{c\nabla}(0) \le \sup_{w^*} \inf_x L(x, w^*) \le \inf_x \sup_{w^*} L(x, w^*) \le P(0).$$

The second part of the statement follows directly from the second part of Corollary 17. □

PROPOSITION 19. $w^* \in \partial_H P(0) \Longrightarrow \inf_x L(x, w^*) = P(0)$. If H is closed under pointwise infimum the converse is true.

Proof. If $w^* \in \partial_H P(0)$, by the Strong Duality Theorem $\phi^c(0,w^*)(0) = P(0)$ and by Proposition 16 and Corollary 15, $\inf_x L(x, w^*) = P(0)$. The converse is obtained applying the second part of Proposition 16 and the Strong Duality Theorem. □

DEFINITION. $(\bar{x}, \bar{w}^*) \in C \times R$ is a saddlepoint of L if $L(\bar{x}, w^*) \le L(\bar{x}, \bar{w}^*) \le L(x, \bar{w}^*)$ for all x, w^* or, equivalently, if $\sup_{w^*} L(\bar{x}, w^*) = \inf_x L(x, \bar{w}^*)$ or, also, if $\sup_{w^*} \inf_x L(x, w^*) = L(\bar{x}, \bar{w}^*) =$
$= \inf_x \sup_{w^*} L(x, w^*)$.

THEOREM 20. $\phi(\bar{x}, 0) = \phi^c(0, \bar{w}^*)(0) \Longrightarrow (\bar{x}, \bar{w}^*)$ is a saddlepoint of L. If ϕ_x is H-convex at 0 and H is closed under pointwise infimum the converse is true. Moreover $L(\bar{x}, \bar{w}^*) = \phi(\bar{x}, 0) = \phi^c(0, \bar{w}^*)(0)$.

Proof. By Propositions 16 and 14 we have :
$\phi^c(0, \bar{w}^*)(0) \le \inf_x L(x, \bar{w}^*) \le L(\bar{x}, \bar{w}^*) \le \sup_{w^*} L(\bar{x}, w^*) \le \phi(\bar{x}, 0)$. Then, $\phi(\bar{x}, 0) = \phi^c(0, \bar{w}^*)(0)$ implies that $\inf_x L(x, \bar{w}^*) = \sup_{w^*} L(\bar{x}, w^*)$, i.e., the condition of saddlepoint for $(\bar{x}, \bar{w}^*)$, and furthermore $L(\bar{x}, \bar{w}^*) = \phi(\bar{x}, 0) = \phi^c(0, \bar{w}^*)(0)$. Conversely, if $(\bar{x}, \bar{w}^*)$ is a saddlepoint of L, ϕ_x is H-convex at 0 and H is closed under pointwise infimum, by Propositions 16 and 14 we conclude that $\phi^c(0, \bar{w})(0) = \inf_x L(x, \bar{w}^*) = \sup_{w^*} L(\bar{x}, w^*) = \phi(\bar{x}, 0)$. □

Let us consider now the case of vertical perturbations, that is, when ϕ is given by

$$\phi(x, w) = f(x) \text{ if } g(x) \ge w, \quad +\infty \text{ otherwise.}$$

Then, if H verifies the next properties a), b), c), the H-lagrangien takes a special feature.

a) H consists of nondecreasing functions,

b) $\forall h \in H \quad t \in R,\ h(t) = \sup\{h'(t) \mid h' \in H,\ h' \leq h,\ h'(0) < +\infty\}$,

c) $h \in H$, $c \in R$ implies $h + c \in H$.

Let $h_o = \inf\{h \mid h \in H,\ h(0) = 0\}$. Then h_o does not necessarily belong to H but it is also a nondecreasing function wich vanishes at 0.

PROPOSITION 21. If a), b), c) hold and ϕ is defined by vertical perturbations, then $L(x, w^*) = f(x) - h_o(<w^*, g(x)>)$ if $w^* \geq 0$, $f(x) - h_o(+\infty)$ if $w^* \geq 0$ (here $h_o(+\infty)$ means $\lim_{t\to+\infty} h_o(t) = \sup_t h_o(t)$ by a)).

Proof. $L(x,w^*) = \Phi_x^c(w^*)(0) = \sup\{h(0) \mid h \in H,\ h(<w^*, w>) \leq f(x)$ if $g(x) \geq w\} = \sup\{h(0) \mid h \in H,\ \sup_{w \leq g(x)} h(<w^*, w>) \leq f(x)\}$.

by a), $\sup_{w \leq g(x)} h(<w^*, w>) = h(t_o)$, being $t_o = <w^*, g(x)>$ if $w^* \geq 0$, $+\infty$ otherwise.

by b), $L(x, w^*) = \sup\{h(0) \mid h \in H,\ h(0) < +\infty,\ h(t_o) \leq f(x)\}$.

By c), $L(x, w^*) = \sup\{c \mid h(t_o) \pm c \leq f(x)$ for some $h \in H$ such that $h(0) = 0\} =$

$$= \sup\{f(x) - h(t_o) \mid h \in H,\ h(0) = 0\} = f(x) - h_o(t_o).$$

By condition a) we have $h_o(<w^*, g(x)>) \leq h_o(\pm\infty)$ $\forall x, w^*$. Hence $L(x, w_1^*) \geq L(x, w_2^*)$ for all x, $w_1^* \geq 0$, $w_2^* \geq 0$. Then, in order to take the supremum of the H-lagrangian with respect to its second variable, we can restrict it to the nonnegative orthant $w^* \geq 0$. The following proposition shows that for a consistent problem with perturbation function being H-convex at 0, each optimal dual solution characterizes the set of optimal primal solutions.

PROPOSITION 22. Let us assume that P is H-convex at 0. Let $\overline{w}^* \geq 0$ be an optimal dual solution and $\overline{x}$ be such that $g(\overline{x}) \geq 0$. Then $\overline{x}$ is an optimal solution if and only if $\overline{x}$ minimizes $f(x) - h(0)(<\overline{w}^*, g(x)>)$ and $h_o(<\overline{w}^*, g(\overline{x})>) = 0$.

Proof. By Proposition 18 and Theorem 11, $\sup_{w^*} \inf_x L(x, w^*) = P(0) = \phi^c(0, w^*)(0)$. If $\overline{x}$ is an optimal solution we have

$$f(\bar{x}) - h_o(<\bar{w}^*,g(\bar{x})>) = L(\bar{x},\bar{w}^*) \le \sup_{w^*} L(\bar{x},w^*) \le \phi(\bar{x},0) = P(0) =$$

$$= \phi^c(0,\bar{w}^*)(0) \le \inf_x L(x,\bar{w}^*) \le f(x) - h_o(<\bar{w}^*,g(x)>) \text{ for any } x.$$

Moreover, by considering $x = \bar{x}$, we get $f(\bar{x}) - h_o(<\bar{w}^*,g(\bar{x})>) = \phi(\bar{x},0) = f(\bar{x})$ and we conclude that $h_o(<\bar{w}^*, g(\bar{x})>) = 0$.

Conversely, if $\bar{x}$ minimizes $f(x) - h_o(<\bar{w}^*, g(x)>)$ and satisfies $h_o(<\bar{w}^*, g(\bar{x})>) = 0$, given any x with $g(x) \ge 0$, we have :

$$f(\bar{x}) = f(\bar{x}) - h_o(<\bar{w}^*,g(\bar{x})>) = \inf_x\{f(x)-h_o(<\bar{w}^*,g(x)>)\} \le$$

$$\le f(x) - h_o(<\bar{w}^*,g(x)>) \le f(x) - h_o(0) = f(x).$$

Hence $\bar{x}$ is optimal solution for the primal problem. □

IV. APPLICATION TO PARTICULAR FAMILIES H

a) H is the family of functions $h_b(t) = t - b$, with $b \in \bar{R}$. In this case the H-convex functions are the closed (l.s.c.) convex functions, $f^c(x^*) = h_{f^*(x^*)}$, f^* being the convex conjugate function ; the H-subgradients are the usual subgradients and the the duality theory and H-lagrangians are the standard ones.

b) H is the family of continuous from the left nondecreasing functions

In [7] Martos defined quasimonotonic functions (over a convex set $X \subset R^n$) as those being both quasiconvex and quasiconcave.
$f : R^n \to \bar{R}$ is said to be quasiaffine if it is quasimonotonic over all R^n.

PROPOSITION 23 (see [8, Theorem 1]. Let $\phi = \{\varphi : R^n \to \bar{R} \mid \exists h \in H,\ x^* \in R^n$ such that $\varphi(x) = h(<x^*, x>)$ for all $x \in R^n\}$. Then the following statements are equivalent :

i) $\varphi \in \phi$, ii) φ is lower semicontinuous quasiaffine,
iii) Every level set of φ, $S(\varphi, \alpha) = \{x \mid \varphi(x) \le \alpha\}$, is a closed halfspace or the empty set or the whole space R^n.

By this proposition, any H-convex function is a supremum of l.s.c. quasiaffine functions, and then it is a l.s.c. quasiconvex function. Before proving the converse of this statement, let us calculate the formula of conjugation to be applied in this case.

PROPOSITION 24. $f^c(x^*)(t) = \inf\{\alpha \mid t \le \sup\{\langle x^*, x\rangle \mid x \in S(f, \alpha)\}$.

Proof. Let $h_o(t) = \inf\{\alpha \mid t \le \sup\{\langle x^*, x\rangle \mid x \in S(f, \alpha)\}$. h_o is a l.s.c. nondecreasing function, that is, $h \in H$. Moreover, for all $x_o \in R^n$

$$h_o(\langle x^*, x_o\rangle) = \inf\{\alpha \mid \langle x^*, x_o\rangle \le \sup\{\langle x^*, x\rangle \mid x \in S(f, \alpha)\}\}.$$

Observing that $\langle x^*, x_o\rangle \le \sup\{\langle x^*, x\rangle \mid x \in S(f, f(x_o)\}$ (since $x_o \in S(f,f(x_o))$) we conclude that $h_o(\langle x^*, x_o\rangle) \le f(x_o)$. Let $h \in H$ be such that $h(\langle x^*, x\rangle) \le f(x)$ for all $x \in R^n$. If $t \le \sup\{\langle x^*, x\rangle \mid x \in S(f, \alpha)\}$ then

$h(t) \le \sup\{h(\langle x^*,x\rangle) \mid x \in S(f,\alpha)\} \le \sup\{f(x) \mid x \in S(f,\alpha)\}=\sup\{f(x) \mid f(x)\le \alpha\le\alpha$.

Hence $h(t) \le \inf\{\alpha \mid t \le \sup\{\langle x^*, x\rangle \mid x \in S(f, \alpha)\} = h_o(t)$. Then we have proved :

$$f^c(x^*) = \sup\{h \mid h \in H,\ h(\langle x^*, x\rangle) \le f(x)\} = h_o. \quad \square$$

Remark. By this proposition $f^c(x^*)(\langle x^*, x_o\rangle)$ coincides with $q(x^*)$, q being the conjugate function of f in x_o defined by Crouzeix [3], His Proposition 7 can be reformulated in the following way :

PROPOSITION 25. $f^{c\nabla}$ is the greatest l.s.c. quasiconvex function majorized by f.

COROLLARY 26. A function is l.s.c. quasiconvex if and only if it is H-convex (i.e., a supremum of l.s.c. quasiaffine functions).

Remark. By Proposition 4 and the previous remark, for the family H of continuous from the left nondecreasing functions $\partial_H f(x_o)$ coincides with $T_f(x_o)$, the tangential of f at x_o defined in [3] . Propositions 5 and 6 in the case of this family H give the first equivalence of Proposition 13'b'), and 13'c'), d') in that paper. Furthermore, the H-duality in this case turns into the second formulation of the dual problem in [2]. The H-lagrangian for a family of perturbed problems becomes :

$$L(x, w^*) = \inf\{\alpha \mid 0 \le \sup\{\langle w^*, w\rangle \mid \phi(x, w) \le \alpha\}.$$

In the case of vertical perturbations all functions ϕ_x are H-convex (l.s.c. quasiconvex). Then the second part of Proposition 14 and Corollary 15 also hold. But H is not closed under pointwise infimum and the remaining propositions about the H-lagrangian cannot fully be applied. Conditions a), b), c) of Proposition 21 hold, and h_o is given by $h_o(t) = -\infty$ if $t < 0$, 0 if $t \geq 0$. Hence we obtain the following formula for the H-lagrangian :

$$L(x, w^*) = f(x) \text{ if } w^* \geq 0 \text{ or if } <w^*, g(x)> \geq 0, +\infty \text{ if } w^* \geq 0 \text{ and } <w^*, g(x)> < 0.$$

This is an extension of the lagrangian mentioned by Luenberger in [6].

c) H is the family of all nondecreasing functions

DEFINITION. A convex set will be called "normal" if it is the intersection of a family of (open or closed) halfspaces.

Remark. Since every closed halfspace is the intersection of all the open halfspaces containing it, we can equivalently define a normal convex set as an intersection of open halfspaces. Every convex set in R is normal but in R^n this is not true. For example, the union of an open halfspace with a point belonging to its border is not a normal convex set.

DEFINITION. A quasiconvex function will be called "normal" if all of its leval sets are normal.

Any quasiconvex function of a single variable is normal, but on R^n this is not true. For example, the indicator function of a not normal convex set is not normal.

The following proposition can be proved similarly to Proposition 23.

PROPOSITION 23'. Let $\Phi = \{\varphi : R^n \to \overline{R} \mid \exists h \in H, x^* \in R^n$ such that $\varphi(x) = h(<x^*, x>) \quad \forall x \in R^n\}$. Then the following statements are equivalent :

i) $\varphi \in \Phi$, ii) φ is normal quasiaffine,

iii) Every level set of φ is an (open or closed) halfspace or ϕ or R^n.

PROPOSITION 24'. $f^c(x^*)(t) = \inf\{f(x) \mid <x^*, x> \geq t\}$.

Proof. Let $h_o(t) = \inf\{f(x) \mid <x^*, x> \geq t\}$. Clearly $h_o \in H$ and verifies : $h_o(<x^*, x_o>) = \inf\{f(x) \mid <x^*, x> \geq <x^*, x_o>\} \leq f(x_o)$ for all $x_o \in R^n$.
If $h \in H$ is such that $h(<x^*, x>) \leq f(x)$ for all $x \in R^n$, since h is nondecrasing we have :

$$<x^*, x> \geq t \Longrightarrow h(t) \leq h(<x^*, x>) \leq f(x), \text{ that is,}$$

$$h(t) \leq \inf\{f(x) \mid <x^*, x> \geq t\} = h_o(t).$$

Then we have proved that $h_o = \sup\{h \mid h \in H,\ h(<x^*, x>) \leq f(x)\ \forall x \in R^n\} = f^c(x^*)$. □

Remark. By this proposition $f^c(x^*)(<x^*, x_o>)$ coincides with $q_G(x^*)$, the G-conjugate of f in x_o mentioned in [3], and $f^{c\nabla}$ is f^{GG}, the quasiconvex biconjugate of f.

PROPOSITION 25'. $f^{c\nabla}$ is the greatest normal quasiconvex function majorized by f.

Proof. First by Proposition 23' $f^{c\nabla}$ is a supremum of functions wich level sets are halfspaces or the empty set or R^n. Hence the level sets of $f^{c\nabla}$ are intersections of halfspaces (normal convex sets) and $f^{c\nabla}$ is normal quasiconvex. Let f_o be any normal quasiconvex function majorized by f. Let $x_o \in R^n$, $\lambda_o < f_o(x_o)$. Then $x_o \notin S(f_o, \lambda_o)$. Since f_o is normal quasiconvex, $S(f_o, \lambda_o)$ is a normal convex set. Then there exists $x_o^* \neq 0$ such that $x \in S(f_o, \lambda_o) \Longrightarrow <x_o^*, x> < <x_o^*, x_o>$. One has :

$$f^{c\nabla}(x_o) = \sup_{x^*} f^c(x^*)(<x^*, x_o>) \geq f^c(x_o^*)(<x_o^*, x_o>)$$

$$= \inf\{f(x) \mid <x_o^*, x> \geq <x_o^*, x_o>\} \geq \inf\{f(x) \mid x \in S(f_o, \lambda_o)\}$$

$$= \inf\{f(x) \mid f_o(x) > \lambda_o\} \geq \inf\{f_o(x) \mid f_o(x) > \lambda_o\} \geq \lambda_o.$$ Hence we obtain

$f^{c\nabla}(x_o) \geq f_o(x_o)$ and the statement is proved. □

COROLLARY 26'. f is H-convex if and only if it is normal quasiconvex.

Remark. By Proposition 4 and the previous remark, for the family H of all nondecreasing functions $\partial_H f(x_o)$ coincides with $\partial^* f(x_o)$, the quasi-subgradient of Greenberg and Pierskalla, mentioned in [3]. Then, Proposition 12 in that paper, stating that $\partial f(x_o) \subset T_f(x_o) \subset \partial^* f(x_o)$ follows directly from our more general Proposition 7. Proposition 4 and 6 for the actual family H give the first equivalence of Proposition 13b) and 13c), d) in that paper. □

Now we can give a counter-example of the first part of Proposition 6. Let $f : R^2 \to R$ be given by $f(x, y) = y^2 \mid (x^2+y^2)$ if $x \leq 0$, $(x, y) \neq (0, 0)$, $1 + x$ if $x \geq 0$, $(x, y) \neq (0, 0)$, 0 if $x = 0$, $y = 0$.
The level sets of this function are:

$$\text{for } \lambda < 0,\ S_\lambda(f) = \phi ;$$

$$\text{for } 0 \leq \lambda < 1,\ S_\lambda(f) = \{(x,y) \mid \sqrt{\lambda}x + \sqrt{1-\lambda}y \leq 0,\ \sqrt{\lambda}x - \sqrt{1-\lambda}y \leq 0,\ x \leq 0\} ;$$

$$\text{for } \lambda \geq 1,\ S_\lambda(f) = \{(x, y) \mid x \geq \lambda-1\}.$$

f is a l.s.c. (then, normal) quasiconvex function because all of its level sets are closed and convex. Indeed, f is continuous at every point different from the origin. Let $x_o = (0, 1)$, $x^* = (\alpha, \beta)$. Following Proposition 24',

$$f^c(x^*)(<x^*, x_o>) = \inf\{f(x, y) \mid \alpha x+\beta\lambda \geq \beta\}.$$

If $\beta > 0$, $f^c(x^*)(<x^*, x_o>) \leq f(-\beta, 1\pm\alpha) = (1+\alpha)^2 \mid (\beta^2+(1+\alpha)^2) < 1.$

If $\beta \leq 0$, $f^c(x^*)(<x^*, x_o>) \leq f(0, 0) = 0 < 1.$

Since $f(x_o) = f(0, 1) = 1$, there is no H-subgradient for f at x_o. The same example can be used in the case of H being the set of continuous from the left nondecreasing functions.

Using the formula of conjugation, the H-lagrangian for a general perturbed problem results :

$$L(x, w^*) = \inf\{\phi(x, w) \mid <w^*, w> \geq 0\} .$$

If ϕ is given by vertical perturbations, all functions ϕ_x are H-convex (normal quasiconvex). Moreover H is closed under pointwise infimum and all propositions about the H-lagrangian can fully be applied. All of the hypothesis of Proposition 21 also hold, and h_o is the same function that in the case before. Then the H-lagrangian is also the same one.

V. POSSIBLE FURTHER GENERALIZATIONS

There are some possible generalizations which perhaps can be useful in nonconvex duality theory :

a) A first generalization can be made by substituting linear functions by more general functions $\omega \in \Omega$. In this case the appropriate definitions of conjugate functions are :

for $f : R^n \to \overline{R}$, $f^c : \Omega \to H$ is given by $f^c(\omega) = \sup\{h \mid h \in H,\ h \circ \omega \leq f\}$;

for $g : \Omega \to H$, $g^c : R^n \to \overline{R}$ is given by $g^c = \sup(g(\omega) \circ \omega)$.

The remaining definitions and propositions about H-conjugates can easily be extended to this more general framework, since linearity is not an essential property. In the case of H being the family of functions h_b with $h_b(t) = t - b$, $b \in \overline{R}$ the the conjugates are equivalent to those appearing in [5]. In that paper Ω-convex sets are defined as the level sets of functions belonging to Ω. Then, defining a function to be Ω-quasiconvex if all of its level sets are Ω-convex and taking H the family of continuous from the left nondecreasing functions, the conjugate function of $f : R^n \to \overline{R}$ is given by $f^c(\omega)(t) = \inf\{\alpha \mid t \leq \sup\{\omega(x) \mid x \in S(f, \alpha)\}\}$ and $f^{c\nabla}$ is the greatest Ω-quasiconvex function majorized by f.

b) Another possible generalization can be obtained by taking Ω a set of functions $\omega : R^n \to C$, with C an arbitrary set and H a family of functions $h : C \to \overline{R}$ being closed under pointwise supremum. The formulae of conjugate functions can be taken as in the last case.

c) Finally, let Ω be an arbitrary set, H a family of maps from Ω into the set of functions from R^n into $\overline{R}$, being closed under pointwise supremum. Then for $f : R^n \to \overline{R}$, $g : \Omega \to H$ we can define :

$$f^c : \Omega \to H,\ f^c(\omega) = \sup\{h \mid h \in H,\ h(\omega) \leq f\} ;$$

$$g^\nabla : R^n \to \overline{R},\ g^\nabla = \sup(g(\omega))(\omega).$$

REFERENCES

1. M. Avriel, Nonlinear programming. Analysis and methods , Prentice-Hall Inc., Englewood Cliffs, N.J. (1976).

2. J.-P. Crouzeix, Polaires quasiconvexes et dualité , C.R. Acad. Sci. Paris (1974) Série A, t. 279, pp. 955-958.

3. J.-P. Crouzeix, Conjugacy in quasiconvex analysis , in Convex Analysis and Its Applications. Proceedings of a Conference Held at Murat-le-Quaire, March 1976, ed. by Auslender. Springer-Verlag, Berlin-Heidelberg (1977).

4. S. Dolecki and S. Kurcyusz, Convexité généralisée et optimisation , C.R. Acad. Sci. Paris, Série A-B, 283 (1976), pp. A-91-94.

5. S. Dolecki and S. Kurcyusz, On Φ-convexity in extremal problems , SIAM J. Control, (1978), Vol. 16, n°2, pp. 277-299.

6. D. G. Luenberger, Quasiconvex Programming , SIAM J. Appl. Math. (1968), 16, pp. 1090-1095.

7. B. Martos, Quasiconvexity and Quasimonotonicity in Nonlinear Programming , Studia Sci. Math. Hung., 2, 265-273 (1967).

8. W. A. Thompson and D.-W. Parke, Generalized Concave Functions , Operations Research, Vol. 21, n°1, pp. 305-313 (1974).

WELL-POSED SADDLE POINT PROBLEMS

Ennio Cavazzuti

University of Modena
Modena, Italy

Jacqueline Morgan

University of Naples
Naples, Italy

We consider the saddle point problem :
To find $(\overline{x}, \overline{y}) \in X \times Y$ such that : $\forall x \in X, y \in Y$

$$f(x, \overline{y}) \leq f(\overline{x}, \overline{y}) \leq f(\overline{x}, y)$$

where X and Y are metric spaces and $f : X \times Y \to \overline{R} = [-\infty, +\infty]$. We introduce the notion of well-posed saddle point problem that is a problem such that any maximinimizing sequence converges to the assumed unique solution of the saddle point problem. We present some characterizations and sufficient conditions for well-posedness in the general case and, in the framework of the theory of duality, necessary and sufficient conditions for the well-posedness of the saddle point problem associated to the Lagrangian. All results refered to Banach (or metric) spaces are given with respect to the strong convergence. Finally we give numerical motivations for the introduction of this notion.

I. INTRODUCTION

Let X and Y be metric spaces and a functional $f : X \times Y \to \overline{R} = [-\infty, +\infty]$.

We consider the following problem :

$$(P)\begin{cases} \text{To find } (\overline{x}, \overline{y}) \in X \times Y \text{ such that} \\ f(x, \overline{y}) \le f(\overline{x}, \overline{y}) \le f(\overline{x}, y) \quad \forall x \in X,\ y \in Y, \end{cases}$$

mainly in connection with approximate solutions of (P). The purpose of this paper is to define the notion of well-posed saddle point problem in analogy with the notion of well-posed minimum problem in the sense of Tickonov [10]. First of all we define maximinimizing sequences and we study their properties. The concept of maximinimizing sequence allows to define, in a natural way, a well-posed saddle point problem as a saddle point problem for which every maximinimizing sequence converges to the unique solution. Therefore, for well-posed saddle point problems, any method generating a maximinimizing sequence is a method for finding approximate solutions of the problem (P). Some examples can be found in [2]. Then we present some characterizations and sufficient conditions for well-posedness. As an application to the duality theory, in the framework of the generalized Lagrangian, we find necessary and sufficient conditions for the well-posedness of the saddle point problem associated with the Lagrangian and these results are subsequently applied to the penalty Lagrangian [9]. Finally we show how to check numerically whether a sequence is maximinimizing or not. This provides a convergence criterion in the case of a well-posed saddle point problem.

II. DEFINITIONS AND PRELIMINARY RESULTS

From now on X and Y are metric spaces, usually equipped with norms. Most of the following results can be extended to a more general context, for example to real locally convex vector spaces or even to topological spaces. Extended real valued functions of the type : $f : X \times Y \to \overline{R} = [-\infty, +\infty]$ are considered ; the values $\pm\infty$ are allowed. In the following y denotes the minimization variable and x the maximization variable. Given $f : X \times Y \to \overline{R}$, we set :

$$(1.1) \qquad h : X \to \overline{R} : h(x) = \inf_{y \in Y} f(x, y) = \inf_{y} f(x, y) \qquad \forall x \in X$$

$$(1.2) \qquad g : Y \to \overline{R} : g(y) = \sup_{x \in X} f(x, y) = \sup_{x} f(x, y) \qquad \forall y \in Y.$$

Then we have : $h(x) \leq f(x, y) \leq g(y) \qquad \forall x \in X, \forall y \in Y.$

DEFINITION 1-1. We say that a function $f : X \times Y \to \overline{R}$ is marginally proper if :

$$\forall x \in X \quad \exists y \in Y : f(x, y) < +\infty \text{ and } \forall y \in Y \quad \exists x \in X : f(x, y) > -\infty.$$

Remark 1-1. If f is a marginally proper function, we cannot have either $h(x) = +\infty$ or $g(y) = -\infty$.

From now on we consider only marginally proper functions. If $U \subset X$, $V \subset Y$ and $f : U \times V \to \overline{R}$, it is usual to extend f to a function on $X \times Y$. The following extensions, called minimal and maximal respectively [8], are commonly known in the literature :

$$f_1(x,y) = -\infty \text{ if } x \notin U, +\infty \text{ if } x \in U \text{ and } y \notin V, f(x, y) \text{ otherwise}$$

$$f_2(x, y) = +\infty \text{ if } y \notin V, -\infty \text{ if } x \in U \text{ and } y \notin V, f(x,y) \text{ otherwise.}$$

Therefore : $h_1(x) = \inf\limits_{y \in Y} f_1(x,y) = -\infty$ if $x \notin U$, $\inf\limits_{y \in V} f(x,y)$ if $x \in U$

$$g_1(x,y) = \sup_{x \in X} f_1(x,y) = +\infty \text{ if } y \notin V, \sup_{x \in U} f(x,y) \text{ if } y \in V,$$

and f_1 is marginally proper if and only if f is marginally proper on $U \times V$. Similarly for f_2. For every marginally proper function f we set :

$$w(x, y) = g(y) - h(x) \geq 0 \qquad \forall (x, y) \in X \times Y. \tag{1.3}$$

This function w is well defined, since additions of the type $(\pm\infty)+(\mp\infty)$ are forbidden.

Let H be an Hilbert space and $\langle ., . \rangle$ a scalar product on H.

Let $f(x, y) = \langle x, y \rangle$, then we have :

$$h(x) = -\infty \text{ if } x \neq 0, 0 \text{ if } x = 0$$

$$g(y) = +\infty \text{ if } y \neq 0, 0 \text{ if } y = 0.$$

Thus : $w(x, y) = +\infty$ if $(x, y) \neq (0, 0)$, if $(x, y) = (0, 0)$.

DEFINITION 1-2. We say that a sequence $(x_n, y_n) \in X \times Y$ is maximinimizing for f (on $X \times Y$) if $\lim_{n\to+\infty} w(x_n, y_n) = 0$.

From now on we write m for minimizing, M for maximizing and M - m for maximinimizing.

Remark 1-2. A function f does not necessarily have M - m sequences even if it is regular and defined on compacts.

Example. Let $f(x, y) = (x+y)^2$ for $x, y \in [-1, +1] = I$. We have : $h(x) = 0 \quad \forall x \in I$, $g(y) = (1+|y|)^2 \quad \forall y \in I$, $w(x, y) = (1+|y|)^2 \quad \forall x \in I, \forall y \in I$ and $\min_{x \in I,\, y \in I} w(x, y) = 1$.

DEFINITION 1-3. We say a function f has the inf-sup property if :

$$\inf_y \sup_x f(x, y) = \sup_x \inf_y f(x, y).$$

Remark 1-3. If f is marginally proper and has an inf-sup value, i.e.,

$$\inf_y \sup_x f(x, y) = \min_y \sup_x f(x, y) = \sup_x f(x, y_o),$$

then $\inf_y \sup_x f(x, y) > -\infty$. In an analogous way, when f is marginally proper and has a sup-inf value, i.e.,

$$\sup_x \inf_y f(x, y) = \max_x \inf_y f(x, y) = \inf_y f(x_o, y),$$

we have $\sup_x \inf_y f(x, y) < +\infty$. Naturally we always have :

$$\inf_y \sup_x f(x, y) \geq \sup_x \inf_y f(x, y).$$

DEFINITION 1-4. We say that $(\overline{x}, \overline{y})$ is a saddle point for f if :

$$f(x, \overline{y}) \leq f(\overline{x}, \overline{y}) \leq f(\overline{x}, y) \qquad \forall x \in X, \forall y \in Y,$$

or equivalently if : $\max_x f(x, \overline{y}) = f(\overline{x}, \overline{y}) = \min_y f(\overline{x}, y)$.

We recall [5] that f has saddle points if and only if :

$$\max_x \inf_y f(x, y) = \min_y \sup_x f(x, y)$$

and any pair $(\bar{x}, \bar{y})$ at which the previous max and min are attained is a saddle point for f. Moreover the set of saddle points for f is a rectangle in $X \times Y$ (that is of the form $U \times V$, $U \subset X$, $V \subset Y$). We remark that if f is marginally proper and has saddle points, then the saddle value is finite.

III. BASIC PROPERTIES OF M - m SEQUENCES

We have previously noticed that not every function has M - m sequences. We now give a characterisation of the functions which do have M-m sequences.

THEOREM 2-1. The following properties are equivalent :

i) f admits M-m sequences ; ii) f has the inf-sup property ;

iii) $\inf_{x,y} w(x, y) = 0$.

Proof. i) $\Longrightarrow$ ii). Since $h(x) = \inf_y f(x, y) \leq \sup_x \inf_y f(x, y) \leq$

$$\leq \inf_y \sup_x f(x, y) \leq \sup_x f(x, y) = g(y),$$

if f admits M-m sequences, we have that

$$0 \leftarrow w(x_n, y_n) = g(y_n) - h(x_n) \leq \inf_y \sup_x f(x,y) - \sup_x \inf_y f(x,y) \leq 0.$$

ii) $\Longrightarrow$ iii). $\inf_{x,y} w(x,y) = \inf_{x,y} (g(y)-h(x)) = \inf_y f(y) - \sup_x h(x) = 0.$ □

iii) $\Longrightarrow$ i). Let (x_n, y_n) be a m sequence for w. Then $w(x_n, y_n) \to \inf_{x,y} w(x, y) = 0$, hence (x_n, y_n) is a M-m sequences for f.

THEOREM 2-2. If (x_n, y_n) is a M-m sequence for f then :

$$f(x_n, y_n) \to \sup_x \inf_y f(x, y) = \inf_y \sup_x f(x, y) = \bar{f}.$$

Proof. We already know that f has the inf-sup property. First of all we suppose that $f(x_n, y_n) \geq \bar{f}$; then it follows :

$$|f(x_n,y_n)-\overline{f}| = f(x_n,y_n) - \overline{f} \le \sup_x f(x,y_n) - \overline{f} = g(y_n) - \sup_x \inf_y f(x,y)$$

$$\le g(y_n) - \inf_y f(x_n,y) = g(y_n) - h(x_n) = w(x_n,y_n) \to 0.$$

The same result follows in the case where $f(x_n, y_n) \le \overline{f}$. □

COROLLARY 2-1. If f admits saddle points, then every M - m sequence (x_n,y_n) is such that $f(x_n, y_n)$ converges to the saddle value.

Now we study the behaviour of the M - m sequences in relation to the minimum and maximum problems on the slices determined by the saddle points of f.

THEOREM 2-3. Let $(\overline{x}, \overline{y})$ be a saddle point for f and let (x_n, y_n) be a M - m for f ; then :

1) (x_n) is a M sequences for the problem $(P_{\overline{y}})$: $\max_x f(x, \overline{y})$;

2) (y_n) is a m-sequence for the problem $(P_{\overline{x}})$: $\min_y f(\overline{x}, y)$.

Proof. 1) Follows from these inequalities :

$$w(x_n,y_n) = \sup_x f(x,y_n) - \inf_y f(x_n,y) \ge \sup_x f(x,y_n) - \max_x \inf_y f(x,y) \ge$$

$$\ge \sup_x f(x,y_n) - f(\overline{x},\overline{y}) \ge f(\overline{x},y_n) - f(\overline{x},\overline{y}) \ge 0. \quad □$$

2) Is obtained similarly.

Remark 2-1. The problems $(P_{\overline{x}})$ and $(P_{\overline{y}})$ can both have more than one solution (so $(P_{\overline{x}})$ and $(P_{\overline{y}})$ are both ill-posed) even in the case when every M - m sequence is convergent to the unique saddle point for f.

Example. $f(x, y) = x.y$ for $x, y \in [-1, +1]$.

THEOREM 2-4. If :

1) f has saddle points,
2) (P_x) is a well-posed minimum problem for all $x \in X$,
3) (P_y) is a well-posed maximum problem for all $y \in Y$,
then :

i) f has unique saddle point,

ii) every M - m sequence for f converges to the unique saddle point for f.

Proof. i) If f has two saddle points $(\overline{x}, \overline{y})$ and (x_1, y_1) we have :

$$f(\overline{x}, \overline{y}) \geq f(x_1, \overline{y}) \geq f(x_1, y_1) \geq f(\overline{x}, y_1) \geq f(\overline{x}, \overline{y})$$

and therefore $f(x_1, \overline{y}) = f(x_1, y_1)$, in contradiction with the fact that (P_{x_1}) is well-posed.

ii) Let $(\overline{x}, \overline{y})$ be the unique saddle point for f. From Théorem 2-3 it follows that if (x_n, y_n) is a M - m sequence for f, then (y_n) is minimizing for $(P_{\overline{x}})$; since $(P_{\overline{x}})$ is assumed to be well-posed we obtain that $y_n \to \overline{y}$. By a similar argument we have $x_n \to \overline{x}$. □

THEOREM 2-5. Let f be such that : $x \to f(x, y)$ and $y \to f(x, y)$ are respectively upper semicontinuous (u.s.c.) for all y and lower semicontinuous (l.s.c.) for all x. Then

i) every limit point of a M - m sequence is a saddle point for f,

ii) if the saddle point for f is unique and if X and Y are compacts, then every M - m sequence converges to the unique saddle point.

Proof. The function $g : Y \to \overline{R}$ defined by $g(y) = \sup_x f(x, y)$ is l.s.c. since it is the supremum of l.s.c. functions. Similarly $h : X \to \overline{R}$ is an u.s.c. function on X.

Let $x_k \to \overline{x}$ and $y_k \to \overline{y}$; then : $g(\overline{y}) \leq \underline{\lim}_k\, g(y_k)$ and $h(\overline{x}) \geq \overline{\lim}_k\, h(x_k)$.

Thus : $0 \leq g(\overline{y}) - h(\overline{x}) \leq \underline{\lim}_k\, g(y_k) - \overline{\lim}_k\, h(x_k) \leq \underline{\lim}_k\, (g(y_k)-h(x_k)) =$

$$= \underline{\lim}_k\; w(x_k,y_k) = 0.$$

It follows that $g(\overline{y}) = h(\overline{x})$ and $(\overline{x}, \overline{y})$ is therefore a saddle point for f. If the saddle point is unique and X and Y are compacts, then obviously every M - m sequences converges to it. □

IV. WELL-POSED SADDLE POINT PROBLEMS

DEFINITION 3-1. We say that f has a well-posed saddle point problem (w.p.s.p.p.) if :

i) f admits a unique saddle point,

ii) every M - m sequence for f converges to the unique saddle point.

THEOREM 3-1. f has a w.p.s.p.p. if and only if the function w defined by (1-3) has 0 as minimum value and the minimum problem for f is well-posed.

Proof. If $(\bar{x}, \bar{y})$ is the saddle point for f we have $w(\bar{x}, \bar{y}) = 0$, hence $(\bar{x}, \bar{y})$ is a minimum point for the nonnegative function w and vice-versa. Every (x_n, y_n) M - m sequence for f is a minimizing sequence for w and $(x_n, y_n) \to (\bar{x}, \bar{y})$, hence the minimum problem for w is well-posed. The converse is also true. □

THEOREM 3-2. f has a well-posed saddle point problem if and only if the following three conditions are satisfied :

i) $\max_x \inf_y f(x, y) = \min_y \sup_x f(x, y)$,

ii) $g(y) = \sup_x f(x, y)$ has well-posed minimum problem,

iii) $h(x) = \inf_y f(x, y)$ has well-posed maximum problem.

Proof. The conditions are necessary. In fact if f has a w.p.s.p.p. then i) is satisfied and there exists a unique saddle point $(\bar{x}, \bar{y})$. Let (x_n) be an M-sequence for h, i.e.,

$$h(x_n) \to \sup_x h(x) = \inf f(\bar{x}, y) = h(\bar{x}) = f(\bar{x}, \bar{y}).$$

Let us show that $x_n \to \bar{x}$. It suffices to verify that $(x_n, \bar{y})$ is M - m for f. In fact $w(x_n, \bar{y}) = g(\bar{y}) - h(x_n) = f(\bar{x},\bar{y}) - h(x_n) = h(\bar{x}) - h(x_n) \to 0$. It follows that $(x_n, \bar{y}) \to (\bar{x}, \bar{y})$ and hence $x_n \to \bar{x}$. Similarly the condition is sufficient.

Let $\bar{y}$ be the unique minimum point for g and $\bar{x}$ the unique max-point for h. From i) we obtain that $g(\bar{y}) = h(\bar{x})$ and $(\bar{x}, \bar{y})$ is the unique saddle point for f.

Let (x_n, y_n) be a M - m sequences for f, we have :

$$0 \le w(x_n,y_n) = g(y_n) - h(x_n) = g(y_n) - g(\bar{y}) + h(\bar{x}) - h(x_n) \to 0.$$

Thus $0 \le g(y_n) - g(\bar{y}) \to 0$, $0 \le h(\bar{x}) - h(x_n) \to 0$, and from ii) and iii) it follows :

$$y_n \to \bar{y}, \quad x_n \to \bar{x},$$

which allows us to conclude that f has a well-posed saddle point problem. □

Remark 3-1. It can happen that f has a well-posed saddle point problem, hence g and h have well-posed minimum problem and well-posed maximum problem respectively, even if both problems $\min_y f(\bar{x}, y)$ and $\max_x f(x, \bar{y})$ are ill-posed.

Example. $f(x, y) = x.y$ for $x, y \in [-1, +1]$. $(0, 0)$ is the unique saddle point for f and $g(y) = |y|$, $h(x) = -|x|$, while $f(0, y) = f(x, 0) = 0$, $\forall x, y \in [-1, +1]$.

Before giving some sufficient conditions for a saddle point problem to be well-posed, we recall some properties about convex functions and existence theorems for a saddle point. If $f_x(y)$, $x \in X$, is a family of quasi-convex (convex), lower semicontinuous functions, then the function $g(y) = \sup_x f_x(y)$ is a quasi-convex (convex) lower semicontinuous (with values in $\bar{R}$). Similarly if $f_y(x)$, $y \in Y$ is a family of quasi-concave (concave) upper semicontinuous functions then $h(x) = \inf_y h_y(x)$ is a quasi-concave upper semicontinuous function.

Theorem (Sion). If f is quasi-concave-convex, upper semicontinuous in x, lower semicontinuous in y, and X and Y are compacts and convex, then f has saddle points. If f is proper the saddle value is finite.

THEOREM [3]. If f is a (quasi) concave-convex, coercive, defined on reflexive Banach spaces, then f has saddle points.

In the following E and F are two reflexive Banach spaces, $X \subset E$ and $Y \subset F$ are convex and closed. We shall use the following definitions :

DEFINITION 3-2. We say that $f : X \to \bar{R}$ is a locally uniformly quasi-concave function [4] if : $\forall x \in X$ there are a function $\delta : t \to \delta(x, t)$ satisfying :

i) $\delta(x, 0) = 0$, $\delta(x, t) > 0$ for $t > 0$,

ii) $t_n \geq 0$, $\delta(x, t_n) \to 0 \Longrightarrow t_n \to 0$,

and a real number $r(x) > 0$ such that :

$$y \in X \cap \text{dom } f \cap B(x,r(x)) \Longrightarrow f(\tfrac{x+y}{2}) \leq \max(f(x),f(y)) - \delta(x, \|y-x\|),$$

where $B(x,r(x)) = \{y \in X : \|y-x\| = r(x)\}$ and $\text{dom } f = \{x \in E : f(x) < +\infty\}$.

Similarly we say that a function f is locally uniformly quasi-concave if (-f) is locally uniformly quasi-convex.

DEFINITION 3-3. We say that a function $f : X \times Y \to \overline{R}$ is locally uniformly quasi-concave-convex if the function $x \to f(x, y)$ and $y \to f(x, y)$ are locally uniformly quasi-convex and locally uniformly quasi-concave respectively for all $(x, y) \in X \times Y$.

THEOREM 3-3. If X and Y are closed and bounded convex and if $f : X \times Y \to \overline{R}$ is a locally uniformly quasi-concave-convex function, u.s.c. in x and l.s.c. in y, then f has a saddle point problem which is well-posed, that is, every M - m sequence strongly converges to the unique saddle point for f.

Proof. The existence of a saddle point follows from Sion's Theorem. The strong convergence of M - m sequences depends on the well-posedness of the problems : $(P_{x_o}) : \min_y f(x_o, y)$ and $(P_{y_o}) : \max_x f(x, y_o)$, $\forall x \in X$ and $\forall y_o \in Y$ (see [4]). □

Remark 3-2. If X and Y are weakly compacts, if f is u.s.c. in x, l.s.c. in y and strictly concave-convex, then the saddle problem defined by f is only weakly well-posed.
Let us consider now the case where constraints are not bounded and let us recall the definition of a coercive function.

DEFINITION 3-4. We say that $f : X \times Y \to \overline{R}$ is a coercive function (semi-compact in Aubin's terminology [1]) if :

$$\exists x_o \in X : \lim_{\substack{y \in Y \\ \|y\| \to +\infty}} f(x_o, y) = +\infty, \quad \exists y_o \in Y : \lim_{\substack{x \in X \\ \|x\| \to +\infty}} f(x, y_o) = -\infty.$$

THEOREM 3-4. Let $X \subset E$, $Y \subset F$ (E and F are reflexive Banach spaces) be closed convex subsets and f a marginally proper, coercive, weakly u.s.c. in x, weakly l.s.c. in x function ; then every M - m sequence is bounded.

Proof. Let (x_n, y_n) be a M - m sequence and let us suppose that (y_n) is unbounded. From the coerciveness hypothesis it follows, subsequencing if necessary, that : $\lim_n f(x_o, y_n) = +\infty$.

From : $f(x_n, y) \geq \inf_y f(x_n, y) = \sup_x f(x, y_n) - w(x_n, y_n) \quad \forall y \in Y$

we deduce : $f(x_n, y_o) \geq f(x_o, y_n) - w(x_n, y_n)$, and finally : $\lim_n f(x_n, y_o) = +\infty$. Hence the sequence (x_n) is bounded as it follows from coerciveness hypothesis. Let us suppose : $x_n \overset{W}{\to} \underline{x}$, then we have :

$$f(\underline{x}, y) \geq \overline{\lim_n} f(x_n, y) \geq \overline{\lim_n} (f(x_o, y_n) - w(x_n, y_n)) = +\infty \quad \forall y$$

and this is in contradiction with the marginally properness hypothesis for f. □

COROLLARY 3-1. If $X \subset E$, $Y \subset F$ are convex closed, if f is coercive, proper strictly quasi-concave-convex and u.s.c. in x, l.s.c. in y, then f defines weakly well-posed saddle point problem.

THEOREM 3-6. If X and Y are convex closed, if f is marginally proper, coercive, locally uniformly quasi-concave-convex, u.s.c. in x, l.s.c. in y, then the saddle point problem defined by f is strongly well-posed.

Remark 3-3. Similar results can be obtained in the case of inequality constraints [4].

V. APPLICATION TO DUALITY THEORY

For the sake of clarity we recall briefly the notion of dual problem and Lagrangian associated to a family of perturbations [8]. Let F be any convex bifunction from R^m to R^n. The (generalized) convex program (P) associated with F is the minimization problem with perturbations in which the function to be minimized over R^n is F_o and the given perturbations are those that replace F_o by F_u for different choices of $u \in R^m$. F_o is the objective function for (P) ; $\inf(F_o)(x)$ is the optimal value in (P) and an element $x \in R^n$ such that $(F_o)(x)$ is finite and equal to the optimal value in (P) is an optimal solution to (P). The perturbation function for (P) is the function in F on R^m given by :

$$(\inf F)(u) = \inf_x (F_u)(x).$$

A vector $u^* \in R^m$ is a Kuhn-Tucker vector for (P) if the quantity : $\inf_{u,x} \{\langle u^*, u\rangle + (F_u)(x)\}$ is finite and equal to the optimal value in (P).

The Lagrangian L of (P), defined on $R^m \times R^n$, is given by :

$$L(u^*, x) = \inf_u \{\langle u^*, u\rangle + (Fu)(x)\}.$$

Then one has the following theorem [8] : If the convex bifunction F is proper and closed, then $\bar{x}$ is an optimal solution and $\bar{u}^*$ a Kuhn-Tucker vector for (P) if and only if $(\bar{u}^*, \bar{x})$ is a saddle point for the Lagrangian L associated with the given family of perturbations. The dual program (P^*) to (P) is the concave program associated with the (concave) adjoint bifunction F^*. The optimal value in (P^*) is $(\sup F^*)(0)$. A Kuhn-Tucker vector for (P^*) is a vector x such that $\sup_{u^*,x^*} (\langle x, x^*\rangle + (F^*x^*)(u^*))$ is finite and equal to the optimal value in (P^*). Moreover if F is closed and proper we have $F^{**} = F$ and the dual program to (P^*) is (P). The convex program (P) is normal if inf F is closed in 0. Many conditions ensure the normality of (P), particularly, when F is a closed convex bifunction, the uniqueness of the Kuhn-Tucker vector ensures the normality of (P). Finally we recall that (P) is strictly consistent if $0 \in \text{int}(\text{dom } F)$ or, which is equivalent, $0 \in \text{int}(\text{dom} \inf_x F)$. We shall use the two following lemmas :

LEMMA 4-1. If F is a convex bifunction from R^m to R^n and (P) is the associated program, then the function $h : R^m \to \bar{R}$ defined by :

$$h(u^*) = \inf_x L(u^*, x)$$

verify : $h(u^*) = (F_o)(u^*)$, that is to say h is the objective function for (P^*) and hence $\sup_{u^*} h(u^*) = \sup_{u^*} \inf_x L(u^*, x)$ is the optimal value in (P^*).

LEMMA 4-2. If F is a closed convex bifunction and (P) is the associated convex program, then the function $g : R^m \to \bar{R}$ defined by : $g(x) = \sup_{u^*} L(u^*, x)$ verify : $g(x) = (F_o)(x)$, that is to say g is g is the objective function for (P) and hence $\inf_x g(x) = \inf_x \sup_{u^*} L(u^*, x)$ is the optimal value in (P).

THEOREM 4-1. Let F be a closed convex bifunction from R^m to R^n and (P) be the associated convex program ; then the two following conditions are equivalent :

i) (u_n^*, x_n) is a M - m sequence for the Lagrangian L,

ii) (P) is normal, the sequence (u_n^*) is maximizing for the objective function of (P^*) and (x_n) is minimizing for the objective function of (P).

Proof. i) $\Longrightarrow$ ii). If (u_n^*, x_n) is a M - m sequence for L, then L has the inf-sup property from the Theorem 2-1, that is :

$$\alpha = \inf_x \sup_{u^*} L(u^*, x) = \sup_{u^*} \inf_x L(u^*, x),$$

hence (P) is normal [8]. Moreover we have :

$$w(u_n^*, x_n) = \sup_u L(u^*, x_n) - \inf L(u_n^*, x) = g(x_n) - h(u_n^*) \to 0,$$

which implies that $g(x_n) \to \alpha$, $h(u_n^*) \to \alpha$.
Finally, recalling that $\alpha = \inf_x g(x) = \sup_{u^*} h(u^*)$, we obtain what is stated in ii).
ii) $\Longrightarrow$ i). If (u_n^*) is maximizing for the objective function of (P^*) and (x_n) is minimizing for the objective function of (P) we obtain, from Lemmas 4-1 and 4-2, that :

$$h(u_n^*) \to \sup_{u^*} h(u^*), \quad g(x_n) \to \inf_x g(x)$$

and since (P) is normal, we have $\sup_{u^*} h(u^*) = \inf_x g(x)$, form which it follows : $w(u_n^*, x_n) = g(x_n) - h(u_n^*) \to 0$.

COROLLARY 4-1. Let f be a closed convex bifunction. The associated Lagrangian has a well-posed saddle point problem if and only if the primal problem (P) and the dual problem (P^*) are well-posed optimization problems.

Of course it is more interesting to find necessary and sufficient conditions which involve only the problem (P). In fact we have :

THEOREM 4-2. Let F be a closed convex bifunction and (P) be the associated convex program. Then the associated Lagrangian defines a well-posed saddle point problem if and only if (P) is well-posed, admits a unique Kuhn-Tucker vector and is strictly consistent.

Proof. The necessity of the condition follows from the previous corollary. To prove the sufficiency of the condition, it is enough to show that $w(u_n^*, x_n) \to 0$ implies $u_n^* \to \bar{u}^*$ and $x_n \to \bar{x}$. Let now $w(u_n, x_n) \to 0$, that is $g(x_n) \to \alpha = \inf \sup L$ and $h(u_n^*) \to \alpha$. From this, (P) being well-posed, it follows $x_n \to \bar{x}$, so it will remain to prove that : $u_n^* \to \bar{u}^*$.

Let $p(u) = \inf_x (F_u)(x)$. If (P) is strictly consistent then p became o proper convex function finite on a neighborhood of 0 and moreover $-h(u^*) = p^*(-u^*)$ [9] (p^* denotes the conjugate of p). From the preceding equality we deduce that every level set $S_\beta = \{u^* : h(u^*) \geq \beta\}$ is compact [8] and hence the sequence (u_n^*) is bounded. Then we can show that every limit point of (u_n^*) is an optimal solution to (P^*), hence a Kuhn-Tucker vector for (P), and from the uniqueness hypothesis we deduce that $(u_n^*) \to \overline{u}^*$. □

Ordinary and penalty Lagrangians cases [9]. We consider the following problem (P) :

$$\text{(P)} \quad \min_{x \in X} f_o(x) \text{ subject to } f_i(x) \leq 0 \qquad i = 1,\ldots,n.$$

For $r \geq 0$ we consider : $L_r(x,y) = \min_{u \in R^n} \{F_r(u,x) + u.y\} \qquad x \in X,\ y \in R^n,$

where $F_r(x,u) = f_o(x) + r \sum_1^n u_i^2$ if $u_i \geq f_i(x)$ $i = 1,\ldots, n$, $+\infty$ otherwise. According to the perturbational theory of duality [8], L_o (which is the ordinary Lagrangian associated with (P)) corresponds to the embedding of (P) in the class of problems where for each $u \in R^n$ the function f_i is perturbed to $f_i - u_i$, $i = 1,\ldots,n$. For $r > 0$ L_r is a Lagrangian in the sense of the theory and it arises in the same way, except that f_o is simultaneously perturbed to $f_o + r u^2$ [9]. From [9] we know that for every $r > 0$ the dual problem (D_r), associated with L_r, has the same solutions as the ordinary dual (D_o). From Theorem 4-2 then have :

COROLLARY 4-2. Suppose the Slater condition is verified in the problem (P), that is : $\exists x : f_i(x) < 0$, $i = 1,\ldots, n$. Then L_r $(r \geq 0)$ has well-posed saddle point problem if and only if (P) is well-posed and has an unique Kunh-Tucker vector.

Remark 4-2. Some conditions for the well-posedness of the problem (P) defined as above are given in [4].

VI. NUMERICAL POINT OF VIEW

Opposit to the notion of minimizing sequence, which uses explicitely the minimum value, the definition of maximinimizing sequence does not refer to the saddle value of the problem. If a well-posed saddle point problem is given, every criterion which ensures that a sequence is a maximinimizing sequence is also a criterion for the convergence of that sequence to the unique saddle point. If we know that a saddle point problem is well-posed and we use an algorithm for which no criterion of convergence is known, we can verify directly the convergence property by means of the maximinimizing property. The same is true if sufficient conditions for the convergence of sequences constructed by means of that algorithm are not verified. Naturally we cannot, in general, determine exactly the quantities $w(x_n, y_n)$ for which we wish to verify the convergence to 0, but only certain approximations of $\sup_x f(x, y_n)$ and $\inf_y f(x_n, y)$. However we can use the following remark.

Remark 5-1. If there is a sequence (t_n, z_n) such that :

i) $f(x_n, z_n) - \inf_y f(x_n, y) \leq \varepsilon_n$ with $\varepsilon_n \to 0$,

ii) $\sup_x f(x, y_n) - f(t_n, y_n) \leq \alpha_n$ with $\alpha_n \to 0$,

iii) $f(x_n, z_n) - f(t_n, y_n) \to 0$ as $n \to +\infty$,

then (x_n, y_n) is a maximinimizing sequence for f.

So, in order to verify numerically that a sequence (x_n, y_n) is maximinimizing, it is possible to use auxiliary sequences (t_n, z_n) obtained by classical methods in optimization and verify whether the quantities $f(x_n, z_n) - f(t_n, y_n)$ converge to 0.

Observation. After the presentation of this paper, the authors have been informed that the concept of maximinimizing sequence had been independently introduced by Mc Linden in a paper entitled "A minimax theorem". Afterwards the authors have received papers [6] and [7] in which the notion of minimaximizing is studied in details and properties of compactness (with respect to the weak topology) are presented in relation with the existence problem of saddle points. So the authors emphasize that the aim of this paper was the study of the strong convergence of the maximinimizing sequences and of the notion of well-posed saddle point problems with respect to the strong convergence.

REFERENCES

1. J.P. Aubin, Mathematical methods of Game and Economic Theory, North Holland, Amsterdam 1979.

2. A. Auslender, Optimisation : Méthodes Numériques, Masson, Paris 1976.

3. H. Brezis, L. Nirenberg and G. Stampacchia, A remark on Ky Fan's minmax principle , Boll Un. Mat. Ital. (Ser. 4) 6, 293-300.

4. E. Cavazzuti and J. Morgan, Problèmes d'optimisation unformément bien posés et méthodes de pénalisation , Boll. Un. Mat. Ital., (to appear).

5. J. Ekeland and R. Temam, Convex analysis and variational problems, Paris Bruxelles, Dunod Gauthiers-Villars (1974).

6. L. Mc Linden, An application of Ekeland's Theorem to minimax theorems , J. Nonlinear Analysis : Theory, Methods and Applications (to appear).

7. L. Mc Linden, A minimax theorem , Math. Operations Research (to appear).

8. R. T. Rockafellar, Convex analysis, Princeton Univ. Press, Princeton, N.J. 1970.

9. R. T. Rockafellar, A dual approach to solving nonlinear programming problems by unconstrainded optimization , Mathematical Programming, 5 (1973), 354-373.

10. A. N. Tickonov, On the stability of the functional optimization problem , USSR Comp. Math. and Math. Phys., vol. 6, n°4 (1966), 631-634.

CONTINUITY PROPERTIES OF PERFORMANCE FUNCTIONS

Jean-Paul Penot

Universite de Pau
Pau, France

Perturbed optimization problems of the following general form

$$(P_w) \quad \text{maximize} \quad f_w(x) \quad \text{over} \quad A_w \subset X$$

where $w \in W$ is a parameter, $f : W \times X \to \overline{\mathbb{R}}$, $f_w(x) = f(w,x)$, can be studied from different points of view. Here we focus our attention on the continuity properties of the performance functional $p : W \to \overline{\mathbb{R}}$ given by

$$p(w) = \sup\{f_w(x) \mid x \in A_w\} \quad ,$$

where W and X are topological spaces and $A : W \rightrightarrows X$ given by $A(w) = A_w$ is a given relation defining the admissible (or feasible) set. The related but more complex problem of establishing continuity properties of the set S_w of solutions of (P_w) will be studied elsewhere. The other aspects of parametric programming (measurability, lipschitzian properties, (generalized) differentiability...) are not considered in this paper although our lecture dealt with the marginal interpretation of multipliers. Let us observe that the generalized subdifferentiability properties of p cannot be deduced from the related properties of the marginal function $m : W \to \overline{\mathbb{R}}$ given by

$$m(w) = \inf\{f_w(x) \mid x \in A_w\}$$

if one uses the subdifferential calculus of [14] or [18] (in contrast with the lipschitzian case studied in [9]). Here of course the studies of m and p are equivalent, changing f into −f . We choose p rather than m for the sake of symmetry of the statements, although m is more closely linked with convex optimization.

We introduce in section 1 new concepts of semicontinuity for multifunctions (or relations). One is a restricted form of upper semicontinuity; the other one enlarges the notion of lower semicontinuity. We relate these two concepts with the definition of compact relation introduced in [7] and [15]. All these notions may be useful elsewhere.

The second section of this paper is devoted to semicontinuity results of the performance function p. The last section deals with a characterization of upper semicontinuity of p in terms of approximate solutions to (P_w).

The results we give could be extended to mappings with values in an ordered topological space using the notions of semicontinuity introduced in [16].

I - CLASSICAL AND NON CLASSICAL CONTINUITY CONCEPTS FOR RELATIONS

In all the sequel we identify a relation (or multifunction) with its graph. The filter of neighborhoods of a point x of a topological space X is denoted by $\mathcal{V}(x)$ or $\mathcal{V}_X(x)$. All topological spaces are supposed to be Hausdorff. If d is a distance on X , and if $B \subset X$, $a \in X$, we set $d(a,B) = \inf\{d(a,b)/b \in B\}$.

We use the classical concepts of semicontinuity for relations as expounded in [3], [11] ; note that they sometimes differ from the earlier definitions given in [1], [2], [10]. In particular a relation $F : X \twoheadrightarrow Y$ between two topological spaces is said to be <u>closed</u> at $a \in X$ if each converging net $((x_i,y_i))_{i\in I}$ of F satisfies $\lim(y_i) \in F(a)$ if $\lim(x_i) = a$. It is said to be <u>upper semicontinuous</u> (u.s.c.) at a if for each open subset V of Y containing F(a) , there exists $U \in \mathcal{V}(a)$ with $F(U) \subset V$; here $F(U) = \bigcup_{x\in U} F(x)$. The following variant was introduced in [16].

1.1. DEFINITION. $F : X \twoheadrightarrow Y$ is graphically upper semicontinuous (g.u.s.c.) at $a \in X$ if for each open subset W of $X \times Y$ containing $\{a\}\times F(a)$, there exists $U \in \mathcal{V}(a)$ with $\{x\} \times F(x) \subset W$ for each $x \in U$,

Hence F is g.u.s.c. at a iff (I_X, F) given by $(I_X , F)(x) = \{x\} \times F(x)$ is u.s.c. at a . However this definition is much more restrictive than mere upper semicontinuity. For instance a constant multifunction may not be g.u.s.c.. A characterization of g.u.s.c. relations will be given later on. For the moment let us note the following elementary result the proof of which is left to the reader.

1.2. PROPOSITION. If F is g.u.s.c. at $a \in X$ then F is u.s.c. at a. The converse is true if $F(a)$ is compact. If X and Y are metrizable, and if $a \in X$ is not an isolated point of X, a closed subset B of Y is compact iff any relation $F : X \rightrightarrows Y$ with $F(a) = B$ is g.u.s.c. at a whenever it is u.s.c. at a.

Observe however that $F : \mathbb{R} \rightrightarrows \mathbb{R}$ given by $F(0) = \mathbb{R}$, $F(x) = [-1,1]$ for $x \neq 0$ is g.u.s.c. at 0 although $F(0)$ is not compact.

We now introduce a weakening of the concept of semicontinuity which will prove very versatile.

1.3. DEFINITION. $F : X \rightrightarrows Y$ is said to be ℓ.s.c. at (a,B) where $a \in X$, $B \subset Y$ if for each net $(x_i)_{i \in I}$ in X with limit a there exist $b \in B$, a subnet $(x_j)_{j \in J}$ of $(x_i)_{i \in I}$, and a net $(y_j)_{j \in J}$ in Y with limit b such that $y_j \in F(x_j)$ for each $j \in J$.

When $B = \{b\}$, F is said to be ℓ.s.c. at (a,b); this definition coincides then with a notion used by several authors. If F is ℓ.s.c. at (a,b) for each $b \in F(a)$, then F is said to be ℓ.s.c. at a; this is the classical definition. Note that this is much more restrictive than mere lower semicontinuity of F at $(a,F(a))$. If nets are replaced by sequences we call F sequentially ℓ.s.c. at (a,B). We observe that in definition 1.3 X could be a subspace of some space X' and $a \in X'$ a limit point of X.

1.4. Examples. a) If F is such that for some $U \in \mathcal{V}_X(a)$ $B = \bigcap_{u \in U} F(u)$ is non void, then F is ℓ.s.c. at (a,B) (and in fact at (a,b) for each $b \in B$).

b) If there exists a finite family $(U_k)_{k \in K}$ such that its union U is in $\mathcal{V}_X(a)$ and $\bigcap_{u \in U_k} F(u) \neq \emptyset$ for each $k \in K$, then F is ℓ.s.c. at $(a,F(a))$. This occurs for instance with $F : \mathbb{R} \rightrightarrows \mathbb{R}$ given by $F(0)=[-1,1]$, $F(x) = \{x\,|x|^{-1}\}$ for $x \in \mathbb{R}\setminus\{0\}$.

c) If there exists $U \in \mathcal{V}_X(a)$ and a relatively compact subset B of Y such that $B \cap F(u) \neq \emptyset$ for each $u \in U$, then F is ℓ.s.c. at $(a,\overline{B})$.

d) Given $A : \mathbb{R}^n \to \mathbb{R}^m$ linear, $b \in \mathbb{R}^m$, let $F : \mathbb{R}^n \twoheadrightarrow \mathbb{R}^n$ be defined as follows : for $w \in \mathbb{R}^n$ $F(w)$ is the set of optimal solutions to

$$(P_w) \quad \text{minimize} \quad (w|x) \quad \text{subjet to} \quad Ax \leq b .$$

If for some $w_o \in \mathbb{R}^n$ $F(w_o)$ is bounded, then F is ℓ.s.c. at $(w_o, F(w_o))$.

1.5. Remark. If Y is a metric space, F is ℓ.s.c. at $(a,b) \in F$ iff there exists $U \in V(a)$ and $f : U \to Y$ with $f(x) \in F(x)$ for each $x \in U$, $f(a) = b$, f being continuous at a (take f such that $d(b,f(x)) \leq 2\, d(b,F(x))$.

The following notion will be crucial for our purposes.

1.6. DEFINITION. [7], [15]. The relation $F : X \twoheadrightarrow Y$ is said to be compact at a if for each net $((x_i,y_i))_{i\in I}$ in F with $\lim x_i = a$, $x_i \neq a$ for each $i \in I$, the net $(y_i)_{i \in I}$ has a cluster point.

The following characterization was obtained in [7] and [15] .

1.7. PROPOSITION. If Y is regular, the relation $F : X \twoheadrightarrow Y$ is compact at a iff there exists a compact subset C of Y such that the relation $\hat{F}$ given by $\hat{F}(a) = C$, $\hat{F}(x) = F(x)$ for $x \neq a$ is u.s.c. at a.

Here C can be taken to be the set of cluster points of the filter basis $\{F(U\setminus\{a\}) \mid U \in V(a)\}$. Note that as we required $x_i \neq a$ in definition 1.6 , $F(a)$ is not necessarily compact (and in fact $F(a)$ is irrelevant) and definition 1.6 is equivalent to definition 3 of [15] with F replaced by its restriction to $X \setminus \{a\}$.

1.8. Remarks. The following assertions are easy to prove.

a) If for some $U \in V(a)$, $F(U\setminus\{a\})$ is relatively compact, then F is compact at a.

b) If F is compact at a, then so is any relation $G : X \twoheadrightarrow Y$ with $G(x) \subset F(x)$ for each $x \in X$.

c) If F is closed at a and if F is compact at a (or more generally if F_a given by $F_a(x) = F(x)\ F(a)$ is compact at a) then F is u.s.c. at a. Conversely, if Y is regular, if F is u.s.c. at a and if $F(a)$ is closed, then F is closed at a and F_a is sequentially compact at a.

d) If F is compact at a and $a \in \operatorname{int} D(F)$ where $D(F)$ is the domain of F then F is ℓ.s.c. at (a,B) for some compact subset B of Y. Conversely, if any relation $G : X \rightarrow Y$ included in F is ℓ.s.c. at (a,Y), then F is compact at a.

The following notion will be a link between the previous definitions.

1.9. DEFINITION. The relation $F : X \rightarrow Y$ is said to be subcompact at a if there exists some relation $G : X \rightarrow Y$ compact at a and $U \in V_X(a)$ with $F(x) \supset G(x) \neq \emptyset$ for each $x \in U_a \cap D(F)$, where $D(F)$ is the domain of F and $U_a = U \setminus \{a\}$.

Obviously, a relation F which is compact at a is subcompact at a ; subcompactness is much less restrictive than compactness. This notion plays a quite important role in analysis (see for instance [12], [17]) in some guise or another.

In contrast with proposition 1.7, in metric spaces subcompactness can be characterized by a lower semicontinuity property.

1.10 PROPOSITION. a) If F is subcompact at a, and if Y is regular, there exists a non void compact subset B of Y such that $F|D(F)$ is ℓ.s.c. at (a,B).

b) If X and Y are metrizable and if for some non void compact subset B of Y the relation $F|D(F)$ is ℓ.s.c. at (a,B), then F is subcompact at a.

Proof a) As the filter base $G(V_X(a))$ is compact by [15] prop.6, its cluster set B is non void and compact by [15] prop.14. Moreover $F|D(F)$ is ℓ.s.c. at (a,B) : for any net $(x_i)_{i \in I}$ with limit a in X there exists a subnet $(x_j)_{j \in J}$ and a converging net $(y_j)_{j \in J}$ with $y_j \in G(x_j)$ for each $j \in J$ whose limit belongs to B.

b) Let us suppose Y is metrizable and X has the following property : there exist a function $f : X \to \mathbb{R}_+$ continuous at a, with $f(a) = 0$, $f(x) > 0$ if $x \in X \setminus \{a\}$. This assumption is satisfied if points are closed in X and a has a countable basis of neighborhood (U_n) (take $f(x) = \inf\{n^{-1} \mid x \in U_k \ \forall k \leq n\})$, in particular if X is metrizable. Let $g(x) = \inf\{d(b,y) \mid b \in B,\ y \in F(x)\}$ for each $x \in X$ and let

$$G(x) = \{y \in F(x) \mid d(y,B) \leq f(x) + g(x)\}$$

Let $(x_i)_{i \in I}$ be a net with limit a in $D(G) \subset D(F)$. Then $(g(x_i)) \to 0$: otherwise there exist a cofinal subset K of I and $\varepsilon > 0$ with

$g(x_k) \geq \varepsilon$ for each $k \in K$; but as there exist $b \in B$, a subnet $(x_j)_{j \in J}$ of $(x_k)_{k \in K}$ and $y_j \in F(x_j)$ with $b = \lim y_j$, we get the contraction $\varepsilon \leq g(x_j) \leq d(y_j,b) \to 0$. Let $z_i \in G(x_i)$ for each $i \in I$. Then $(d(z_i,B)) \to 0$ and there exists a net $(b_i)_{i \in I}$ in B with $(d(b_i,z_i)) \to 0$. As B is compact, (b_i), hence (z_i) has a cluster point. □
Now let us study the relationships between graphical upper semi-continuity of F at a and compactness of F at a .

1.11 PROPOSITION. a) If F is compact at a and closed at a , then F is g.u.s.c. at a .
b) If F is g.u.s.c. at a then F is sequentially compact at a ; moreover, if $F(a)$ is closed and Y is regular, F is closed at a .

Proof. a) If F is not g.u.s.c. at a there exist an open subset W of $X \times Y$ containing $\{a\}\times F(a)$, a net $(x_i)_{i \in I}$ with limit a in X and for each $i \in I$ $y_i \in F(x_i)$ with $(x_i,y_i) \notin W$. Thus $x_i \neq a$ and $(y_i)_{i \in I}$ has a cluster point b . As F is closed at a , $b \in F(a)$
On the other hand, we must have $(a,b) \in X \times Y \setminus W$, a contraction.
b) Suppose that some sequence $((x_n,y_n))$ of F is such that (x_n) converges to a , $x_n \neq a$ for $n \geq 1$ and (y_n) has no cluster point. Let $M = \{(x_n,y_n) | n \geq 1\}$ and let $W = X \times Y \setminus M$; W is an open subset of $X \times Y$. If we show that W contains $\{a\}\times F(a)$ we get a contradiction with F g.u.s.c. Let $b \in F(a)$. There exist $V \in \mathcal{V}(b)$ and $k \in \mathbb{N}$ with $y_n \notin V$ for $n > k$. Let $U = X\setminus\{x_n | n \leq k\}$. Then $U \times V \cap M = \emptyset$ and $(a,b) \in W$. The last assertion is valid whenever F is u.s.c. at a , hence a fortiori if F is g.u.s.c. at a . □

Thus, when X is matrizable and F is closed at a, the relation F is compact at a iff F is g.u.s.c. at a.

1.12 PROPOSITION. Let $F : X \twoheadrightarrow Y$ be a relation with Y metrizable.
a) If X is first countable at a and if F is sequentially compact at a there exists $U \in \mathcal{V}(a)$ such that $F(U_a)$ is bounded, with $U_a = U\setminus\{a\}$.
b) If the bounded subsets of Y are relatively compact and if for some $U \in \mathcal{V}(a)$ $F(U_a)$ is bounded, then F is compact at a .

Proof a) Let $b \in Y$ arbitrary and let $(U^n)_{n \geq 1}$ be a countable basis of $\mathcal{V}(a)$. If for each $n \geq 1$ $F(U_a^n)$ is not bounded, there exists a sequence (x_n) in $X \setminus \{a\}$ with $x_n \in U^n$ for each n and a sequence (y_n) in Y with $y_n \in F(x_n)$ and $d(b,y_n) \geq n$ for each n. Then (y_n) has no cluster point and F is not sequentially compact at a.

b) Let $(x_i)_{i\in I}$ be a net in $X \setminus \{a\}$ with limit a and let $(y_i)_{i\in I}$ be such that $y_i \in F(x_i)$ for each $i \in I$. There exists some $k \in I$ such that $x_i \in U$ for $i \geq k$. Then $(y_i)_{i\geq k}$ is contained in the relatively compact subset $F(U_a)$ of Y, hence has a cluster point. □

II - SEMICONTINUITY OF THE PERFORMANCE FUNCTIONAL

In the sequel W and X are topological spaces and 0 is a particular point of W. Given $f : W \times X \to \overline{\mathbb{R}} = [-\infty,+\infty]$ and a relation $A : W \rightarrow\!\!\!\!\!\!\triangleright X$ we study the performance functional

$$p(w) = \sup\{f(w,x) \mid x \in A(w)\} .$$

The following results are easy modifications of classical facts. However, the formulation we give enables one to combine them with the criteria of the preceding paragraph.

2.1. PROPOSITION. If A is g.u.s.c. at 0 and if f is u.s.c. at each point of $\{0\}\times A(0)$, then p is u.s.c. at 0.

Proof. The result is obvious if $p(0) = +\infty$. Suppose $p(0) < +\infty$ and let $r > p(0)$. Then $B = f^{-1}([-\infty,r[)$ is a neighbourhood of $\{0\}\times A(0)$. As A is g.u.s.c. at 0 we can find $U \in \mathcal{V}(0)$ such that $A \cap U\times X \subset B$. Thus for each $w \in U$ we have

$$p(w)=\sup\{f(w,x)\,|\,x\in A(w)\} \leq \sup\{f(w,x)\,|\,(w,x)\in B\} \leq r . \quad \square$$

Combining propositions 1.2 and 2.1 we recover the well-known fact that p is u.s.c. at 0 if f is u.s.c., A is u.s.c. at 0 and $A(0)$ is compact.

2.2. PROPOSITION. If 0 belongs to the interior of the domain $D(A)$ of A, A is ℓ.s.c. at $(0,B)$ with $B \neq \emptyset$, $B \subset S(0)=\{x\in X\,|\,f(0,x)=p(0)\}$ and if f is ℓ.s.c. at each point $(0,b)$ with $b \in B$, then p is ℓ.s.c. at 0.

Proof. Let $(w_i)_{i\in I}$ be a net with limit 0 in W. As 0 is interior to $D(A)$ we can find a subnet $(w_j)_{j\in J}$ and a net $(x_j)_{j\in J}$

with $x_j \in A(w_j)$ for each $j \in J$ and $\lim_{j\in J} f(w_j,x_j) = \liminf_{i\in I} p(w_i)$.
Taking another subnet if necessary, we can suppose $(x_j)_{j\in J}$ converges to a point b in B. Then, as $b \in S(0)$ we have

$$\liminf_i p(w_i) = \lim_j f(w_j,x_j) \geq f(0,b) = p(0). \quad \square$$

The following variant does not suppose (P_0) has optimal solutions.

2.3. PROPOSITION. ([1]) If A is ℓ.s.c. at 0 and if f is ℓ.s.c. at each point of $\{0\} \times A(0)$, then p is ℓ.s.c. at 0.

Proof. If $A(0) = \emptyset$ then $p(0) = -\infty$ and the result is obvious. Let us suppose $p(0) > -\infty$ and let $r < p(0)$. Then there exists $b \in A(0)$ with $f(0,b) > r$ and we can find $Z \in \mathcal{V}_W(0)$, $V \in \mathcal{V}_X(b)$ with $Z \times V \subset f^{-1}((r,+\infty))$. As A is ℓ.s.c. at $(0,b)$, shrinking Z if necessary, we can suppose $A(z) \cap V \neq \emptyset$ for each $z \in Z$. Then for each $z \in Z$ we can find $x \in A(z) \cap V$ so that $p(z) \geq f(z,x) > r$. $\square$

Now we introduce a kind of coerciveness condition. For $r \in \overline{\mathbb{R}}$ we set $F_r = \{(w,s) \mid f(w,x) > r\}$, $F_r(w) = \{x \mid (w,x) \in F_r\}$.

2.4. PROPOSITION. Suppose A is closed at 0, f is u.s.c. at each point of $\{0\}\times A(0)$ and for each $r > p(0)$ the relation $A_r : w \to A(w) \cap F_r(w)$ is ℓ.s.c. on its domain at $(0,X)$, or subcompact at 0. Then p is u.s.c. at 0.

Proof. It suffices to consider the case in which A_r is ℓ.s.c. at $(0,X)$ on its domain. Suppose there exists a net $(w_i)_{i\in I}$ with limit 0 in X such that $\limsup_{i\in I} p(w_i) > p(0)$. Let $r \in (p(0), \limsup_{i\in I} p(w_i))$ and let $(w_j)_{j\in J}$ be a subnet with $\lim_{j\in J} p(w_j) = \limsup_{i\in I} p(w_i)$, $p(w_j) > r$ for each $j \in J$. Thus for each $j \in J$ $A(w_j) \cap F_r(w_j) \neq \emptyset$ and $w_j \in D(A_r)$. Then there exists a subnet $(w_k)_{k\in K}$ of $(w_j)_{j\in J}$, and a converging net $(x_k)_{k\in K}$ with $x_k \in A_r(w_k)$ for each $k \in K$.
As A is closed at $0, x := \lim x_k$ belongs to $A(0)$. Then

$$f(0,x) \geq \limsup_{k\in K} f(w_k,x_k) \geq r > p(0),$$

a contradiction. $\square$

2.5. Remarks a) If F_r is compact at 0 then A_r is subcompact at 0. This is the case if $f(w,x) = f_o(x)$ and f_o is sup-compact (i.e. for each $r \in \mathbb{R}$, $f_o^{-1}([r,+\infty])$ is compact). This also occurs if X is finite dimensional and $\lim_{\substack{w \to 0 \\ |x| \to +\infty}} f(w,x) = -\infty$.

b) If the following assumption (T) is satisfied then for each $r > p(0)$ the relation A_r is subcompact at 0:

(T) for each $r > p(0)$ there exists a compact subset K_r of X and a neighbourhood W_r of 0 in W such that $A_r(w) \cap K_r \neq \emptyset$ if $w \in D(A_r) \cap W_r$.

In fact $G_r = A_r \cap W \times K_r$ is a compact selection of A_r. Let us observe that condition (T) is weaker than the tameness condition used by R.T. Rockafellar in [19] for $X = \mathbb{R}^n$, which is as follows : there exists a compact subset C of X such that for each $\varepsilon > 0$ there exists $Z_\varepsilon \in \mathcal{V}_W(0)$ and $r_\varepsilon < p(0)$ with

$$p(w) = \sup\{f(w,x) \mid d(x,C) \leq \varepsilon, \quad x \in A(w)\}$$

whenever $w \in Z_\varepsilon$ satisfies $p(w) > r_\varepsilon$(if this last requirement is dropped, one has (H_1) of [9]). In fact, given $r > p(0)$ we take ε arbitrary, for instance $\varepsilon = 1$, in the tameness condition and we set

$$K_r = \{x \in X \mid d(x,C) \leq 1\} \quad , \quad W_r = Z_1 .$$

Then, if $w \in D(A_r) \cap W_r$ we have $p(w) > r > r_1$, so that

$$\sup\{f(w,x) \mid x \in A(w) \cap K_r\} = p(w) > r$$

and $A(w) \cap K_r \neq \emptyset$.

Correspondingly, our conclusion is weaker than the strict upper semicontinuity of p at 0 given in [19].

c) Our condition is more general than the assumption made in [20], lemma 2.2, that the mapping $q : X \to \overline{\mathbb{R}}$ given by $q(x) = \sup_{w \in W} f(w,s)$ is sup-relatively compact (i.e. for each $r \in \mathbb{R}$ $q^{-1}([r,+\infty])$ is relatively compact). In fact for each $w \in W$ we have

$$F_r(w) \subset q^{-1}([r,+\infty]),$$

so that F_r is compact at each w_o in W. □

We consider now a mathematical programming problem in which the constraints are given in terms of relations. We suppose W, X, Z are topological spaces, B, C, G are subsets of $W \times X$, $W \times Z$, $X \times Z$ respectively and we set

$$A = \{(w,x) \in W \times X \mid (w,x) \in B, \ \exists z \in Z, \ (w,z) \in C, (x,z) \in G\}$$

Viewing A, B, C, G as multifunctions we have

$$A(w) = B(w) \cap G^{-1}(C(w))$$

2.6 PROPOSITION. Suppose $f_o : X \to \overline{\mathbb{R}}$ is u.s.c., $f : W \times X \to \overline{\mathbb{R}}$ is given by $f(w,x) = f_o(x)$, $C : W \rightrightarrows X$ is closed and compact, $(B^{-1},G) : X \rightrightarrows W \times Z$ maps closed subsets of X onto closed subsets of $W \times Z$. Then p is u.s.c. on W.

Let us observe that when B is a constant relation, the requirement on (B^{-1},G) reduces to the same condition on G. Conditions ensuring that a relation maps closed subsets onto closed subsets are given in [4], [5], [6], [8], [13], [17].

Proof. Let $(w_i)_{i \in I}$ be a net with limit w_o in W and let $r < \limsup_{i \in I} p(w_i)$. We have to show that $p(w_o) \geq r$. Without loss of generality we may suppose $p(w_i) > r$ for each $i \in I$, so that we can find $x_i \in A(w_i)$ with $f(w_i,x_i) > r$ and $z_i \in G(x_i) \cap C(w_i)$. As C is closed and compact, a subnet $(z_j)_{j \in J}$ of $(z_i)_{i \in I}$ has a limit $z_o \in C(w_o)$. As the set $X_r = f_o^{-1}([r,+\infty])$ is closed in X, its image by (B^{-1},G) is closed in $W \times Z$, so that (w_o,z_o) belongs to $(B^{-1},G)(X_r)$. Thus there exists $x_o \in X_r$ with $w_o \in B^{-1}(x_o)$, $z_o \in G(x_o)$, so that $x_o \in A(w_o)$ and $p(w_o) \geq f(w_o,x_o) = f_o(x_o) \geq r$. □

III- APPROXIMATE SOLUTIONS AND UPPER SEMICONTINUITY OF THE PERFORMANCE

In order to avoid the assumption that the problem (P_w) has optimal solutions we are led to introduce the set of approximate solutions of (P_w) : for $\varepsilon > 0$ it is

$$S_\varepsilon(w) = \{x \in A(w) \mid f(w,x) \geq p(w)-\varepsilon\} \quad \text{if} \quad p(w) < +\infty$$

$$S_\varepsilon(w) = \{x \in A(w) \mid f(w,x) \geq \varepsilon^{-1}\} \quad \text{if} \quad p(w) = +\infty$$

We denote by $\tilde{S} : \mathring{\mathbb{R}}_+ \times W \rightrightarrows X$, with $\mathring{\mathbb{R}}_+ = (0,+\infty)$, the relation given by $\tilde{S}(r,w) = S_r(w)$. We observe that $S_r(w) \neq \emptyset$ if $r > 0$ and $A(w) \neq \emptyset$.

It will be convenient to call a subset T of $\mathring{\mathbb{R}}_+ \times W$ a tube around 0 if for each $r > 0$ $T(r) := \{w \in W \mid (r,w) \in T\}$ is a neighbourhood of 0 in W.

3.1 DEFINITION. ([15]) The perturbed problem (P_w) is said to be well-conditioned at 0 if the relation $\tilde{S} : \mathring{\mathbb{R}}_+ \times W \rightrightarrows X$ is compact at $(0,0)$. It is said to be (sequentially) well-set at 0 if $\tilde{S}|D(\tilde{S})$ is (sequentially) ℓ.s.c. at $(0,0,X)$.

We can extend $\tilde{S}$ to $\mathbb{R}_+ \times W$ by setting $\tilde{S}(0,w) = S(w)$ but requiring this extension to be compact or (sequentially) ℓ.s.c.at $(0,0,X)$ would be a stronger assumption we do not wish to make. Of course (P_w) is well-set at 0 if it is well-conditioned at 0.

3.2. PROPOSITION. Suppose the perturbed problem (P_w) is (sequentially) well-set at 0 and

a) A is (sequentially) closed at 0,

b) f is (sequentially) u.s.c. at each point of $\{0\}\times A(0)$.

Then p is (sequentially) u.s.c. at 0. Moreover, if $A(0)$ is non empty (P_0) has an optimal solution.

Proof. Let (w_n) a sequence in W with limit 0 such that $p(w_n)$ has a limit r in $\overline{\mathbb{R}}$. Let us show $r \le p(0)$. This is obvious if $p(0)=+\infty$ or if $r = -\infty$, so that we can suppose $p(w_n) > -\infty$ for each n. Let (ε_n) be any sequence in $(0,+\infty)$ with limit 0. As $A(w_n) \neq \emptyset$ since $p(w_n) \neq -\infty$, we have $\tilde{S}(\varepsilon_n, w_n) \neq \emptyset$. As (P_w) is sequentially well-set, we can find an infinite subset N of $\mathbb{N}$ and a sequence $(x_n)_{n\in N}$ with limit x in X and $x_n \in \tilde{S}(\varepsilon_n,w_n)$ for each $n\in N$. By our assumption on A, $x \in A(0)$. Moreover, when $p(w_n) < +\infty$ for each n in an infinite subset M of N we have

$$p(0) \ge f(0,x) \ge \limsup_{m\in M} f(w_m,x_m) \ge \limsup_{m} (p(w_m)-\varepsilon_m) = r$$

whereas, when $p(w_k) = +\infty$ for an infinite subset K of N we have

$$p(0) \ge f(0,x) \ge \limsup_{k\in K} f(w_k,x_k) \ge \limsup_{k\in K} \varepsilon_k^{-1} = +\infty \ge r .$$

If $A(0)$ is non empty, we can choose $w_n = 0$ in what precedes, so that we get $x \in A(0)$ with $f(0,x) = p(0)$, i.e. $x \in S(0)$.

The proof in the non sequential case is similar : starting with a net $(w_i)_{i\in I}$ converging to 0 we get a converging net $(x_j)_{j\in J}$ in X with $x_j \in \tilde{S}(\varepsilon_{n(j)}, w_{i(j)})$, where (ε_n) is any sequence with limit 0 in $(0,+\infty)$. Then if $K = \{j \in J / p(w_{i(j)}) = +\infty\}$ is cofinal we have $p(0) = +\infty$ as above, whereas if there exists $j_o \in J$ with $j \notin K$ for each $j \ge j_o$ we have $p(0) \ge \limsup_{j\ge j_o} (p(w_{i(j)})-\varepsilon_{n(j)}) = r$ as above. □

3.3. THEOREM.

1) Suppose the restriction $\tilde{S}|T$ of $\tilde{S}$ to some tube T around 0 is subcompact at $(0,0)$, A is closed at 0 and f is u.s.c. at each point of $\{0\}\times A(0)$.

Then p is u.s.c. at 0. Moreover, if $A(0)$ is non empty,

(P_o) has an optimal solution.

2) Conversely, if p is finite and u.s.c. at 0, if f and A are ℓ.s.c. at some point $(0,a)$ with $a \in S(0)$, and if X is first countable at a, then there exists a tube T around 0 in $(0,+\infty) \times W$ such that $\tilde{S}|T$ is subcompact at 0.

Proof. 1) The proof is similar to the preceding one. Let $(w_i)_{i \in I}$ be a net in W with $\lim w_i = 0$, $\lim p(w_i) = r$; let us show that $r \le p(0)$. We can suppose $p(w_i) > -\infty$ for each $i \in I$, so that for any sequence (ε_n) in $(0,+\infty)$ with limit 0, the set $\tilde{S}(\varepsilon_n, w_i)$ is non void. For each $n \in \mathbb{N}$ we choose $i_n \in I$ such that $w_i \in T(\varepsilon_n)$ for $i \ge i_n$. Let

$$J = \{(n,i) \in \mathbb{N} \times I \mid i \ge i_n\}, \quad z_j = (\varepsilon_n, w_i) \quad \text{for} \quad j = (n,i).$$

Then $(z_j)_{j \in J}$ is a net in T with limit $(0,0)$. Let $\tilde{S}_o : T \rightrightarrows X$ be a selection of $\tilde{S}(\tilde{S}_o \subset \tilde{S})$ with domain $D(\tilde{S}) \cap T$, compact at $(0,0)$. Let $x_j \in \tilde{S}_o(z_j)$ for $j \in J$ (recall $z_j \in D(\tilde{S})$). The net $(x_j)_{j \in J}$ has a cluster point $x \in A(0)$ as $(w_{i(j)})_{j \in J} \to 0$ if $j = (n(j), i(j))$. As $\lim_{j \in J} (\varepsilon_{n(j)}) = 0$, estimates as in the proof of proposition show that $r \le p(0)$. The proof of the second assertion is the same as above.

2) As p is u.s.c. at 0, for each $n \in \mathbb{N}$

$$W_n = \{w \in W \mid p(w) < p(0) + 2^{-n}\}$$

is a neighbourhood of 0 in W. Let (U_n) be a countable basis of $\mathcal{V}_X(a)$ with $U_{n+1} \subset U_n$ for each $n \in \mathbb{N}$. As f is ℓ.s.c. at $(0,a)$ we can find a mapping $k : \mathbb{N} \to \mathbb{N}$ with $k(n) \ge n$ for each $n \in \mathbb{N}$ and a sequence (W'_n) in $\mathcal{V}_W(0)$ such that

$$f(w,x) \ge f(0,a) - 2^{-n} \quad \text{for} \quad (w,x) \in W'_n \times U_{k(n)}.$$

As A is ℓ.s.c. at $(0,a)$ we can choose $W''_n \in \mathcal{V}_W(0)$ so that

$$A(w) \cap U_{k(n)} \ne \emptyset \quad \text{for each} \quad w \in W''_n.$$

Let $T(r) = W_n \cap W'_n \cap W''_n$ for $r \in (\frac{1}{n+1}, \frac{1}{n}]$, and let $T = \bigcup_{r>0} \{r\} \times T(r)$. It is a tube around 0 and for each $(r,w) \in T$ we can find

$$x(r,w) \in A(w) \cap U_{k(n)}$$

where $n \in \mathbb{N}$ is given by $n \le r^{-1} < n+1$, so that

$$f(w, x(r,w)) \ge f(0,a) - 2^{-n} = p(0) - 2^{-n} > p(w) - 2^{-n+1}$$

as $w \in W_n \cap W'_n$. Then $\tilde{S}_o(r,w) = \{x(r,w)\}$ is a selection of $\tilde{S}|T$ with domain T and $\tilde{S}_o$ is compact at $(0,0)$ as $x(r,w)$ converges to a as (r,w) converges to $(0,0)$ since $x(r,w) \in U_{k(n)} \subset U_n$ if $r^{-1} \in [n, n+1)$. □

An inspection of the preceding proofs yields the following result.

3.4. COROLLARY. Suppose X is first countable, A is continuous at 0, with $S(0) \neq \emptyset$, f is finite and continuous at each point of $\{0\}\times A(0)$. Then the following assertions are equivalent

a) p is u.s.c. at 0 ;

b) there is some tube T around 0 such that $\tilde{S}|T$ is subcompact at $(0,0)$;

c) there is some tube T around 0 such that $\tilde{S}|T$ is ℓ.s.c. at $(0,0,X)$;

d) there is some tube T around 0 and a mapping $s : T\cup\{(0,0)\} \to X$ continuous at $(0,0)$ with $s(r,w) \in \tilde{S}(r,w)$ for each $(r,w) \in T$.

REFERENCES

1. C. Berge, Espaces topologiques, fonctions multivoques, 2nd edition Dunod, Paris, 1966. English translation, Topological Spaces, Macmillan, New York, 1963.

2. G. Choquet, Convergences. Ann. Univ. Grenoble 23 (1947) 55-112.

3. J.P. Delahaye, J. Denel, The continuities of the point-to-set maps, definitions and equivalences. Math. Prog. Study 10 (1970), 8-12.

4. J.P. Dedieu, Critères de fermeture pour l'image d'un fermé non convexe par une multiapplication. C.R. Acad. Sc. Paris 187 (1978) 941-943.

5. J.P. Dedieu, Cônes asymptotes d'ensembles non convexes. Applications à l'optimisation in Analyse non convexe , Pau (1977), Bull. Soc. Math. de France, Mémoire n°60 (1979) 31-44.

6. J. Dieudonné, Sur la séparation des ensembles convexes. Math. Annalen 163 (1966) 1-3.

7. S. Dolecki, A. Lechicki, Semi-continuité supérieure forte et filtres adhérents. C.R. Acad. Sc. Paris, 293 (1981) 219-221.

8. J. Gwinner, Closed images of convex multivalued mappings in linear topological spaces with applications. J. Math. Anal. Appl. 60 (1) (1977) 75-86.

9. J.B. Hiriart-Urruty, Gradients généralisés de fonctions marginales SIAM J. Control Optimization 16 (1978) 301-316.

10. W.W. Hogan, Point-to-set maps in mathematical programming. SIAM Review 15 (3) (1973) 591-603.

11. P. Huard, Background to point-to-set maps in mathematical programming Math. Progr. Study 10 (1979) 1-7 and 14-28.

12. P. Lévine, J. Ch. Pomerol, Sufficient conditions for Kuhn-Tucker vectors in convex programming. SIAM J. Control 17 (1979) 565-567.

13. J.P. Penot, Critères de fermeture et applications.Mimeographied notes, Pau Univ.

14. J.P. Penot, Calcul sous-différentiel et optimisation. J. Funct. Anal. 27 (2) (1978) 248-276.

15. J.P. Penot, Compact nets, filters and relations (to appear in J. Math. Anal. and Appl.).

16. J.P. Penot, M. Théra, Semi-continuous mappings in general topology Arch. Math. 38 (2) (1982) 158-166.

17. J. Ch. Pomerol, Contribution à la programmation mathématique : existence de multiplicateurs de Lagrange et stabilité. Thèse, Univ. Paris VI, 1980.

18. R.T. Rockafellar, Generalized directional derivatives and subgradients of nonconvex functions. Canad. J. Math. 32 (1980) 257-280.

19. R.T. Rockafellar, Lagrange multipliers and subderivatives of optimal value functions in nonlinear programming. Math. Programming Study 17 (1982) 28-66.

20. I. Singer, Duality theorems for linear systems and convex systems. J. Math. Anal. Appl.76 (2) (1980) 339 368.

ON A GENERAL FORMULATION OF THE HAHN-BANACH PRINCIPLE WITH APPLICATION TO OPTIMIZATION THEORY

W. Oettli

Universität Mannheim
Mannheim, West Germany

ABSTRACT

Conditions are given which ensure that an extended real-valued function admits a support function which is either real-valued and linear or extended real-valued and simultaneously convex and concave. The result is applied to Kuhn-Tucker theory.

I. INTRODUCTION

Starting from recent work by König and Fuchssteiner [1], [2], where the usual convexity requirements of the Hahn-Banach theorem have been widely relaxed, we presented in [3] a result about the existence of a linear support function for an extended real-valued function f defined on a real vector space. Among the assumptions of this result is a "regularity" assumption which ensures that the origin is an internal point of the domain of f, hence in particular $f(0) < +\infty$. In many applications it is desirable to avoid such a regularity assumption and to admit even the possibility that $f(0) = +\infty$. Below we show that this can be done at the expense that the resulting support function is no longer linear, but extended real-valued and simultaneously convex and concave. With an added regularity assumption the resulting function is immediately seen to be again linear. The implications of this result for the Kuhn-Tucker theory of convex optimization are twofold. First we may relax the convexity assumption. Second we may drop the customary regularity assumption, provided we admit extended real-valued multiplier functions. The latter result has already been pointed out for the convex case in [4], where it was derived from a lemma of Stone [5]. The present version is such that Stone's lemma turns out as a special case.

Let us here briefly recall some terminology. We denote the extended reals by $\overline{\mathbb{R}} := \mathbb{R} \cup \{+\infty, -\infty\}$. In dealing with $\overline{\mathbb{R}}$ we agree that $0 \cdot (+\infty) = 0 \cdot (-\infty) = 0$. Let E be a real vector space. A function $g : E \to \overline{\mathbb{R}}$ is called positively homogeneous iff for all $x \in E$ holds $g(\lambda x) = \lambda g(x)$ $\forall \lambda \geq 0$ (implying, in particular, $g(0) = 0$). A function $g : E \to \overline{\mathbb{R}}$ is called convex iff

$$g(\lambda x + (1-\lambda)y) \leq \lambda g(x) + (1-\lambda) g(y)$$

for all x,y satisfying $g(x) < +\infty$, $g(y) < +\infty$, and for all $\lambda \in [0,1]$. g is called concave iff $-g$ is convex. A function which is simultaneously convex and concave and real-valued on all of E is affine-linear in the usual sense. If $C \subset E$, then by cone conv C (the conical convex hull of C) we denote the collection of all nonnegative linear combinations formed from elements of C. Finally we adopt the convention that $\inf \{g(x) \mid x \in \emptyset\} = +\infty$, $\sup \{g(x) \mid x \in \emptyset\} = -\infty$.

II. MAIN RESULT

The Hahn-Banach theorem has many equivalent formulations. One of them is that a convex real-valued function f with $f(0) = 0$ admits a linear supporting function at the origin. This can be extended as follows.

THEOREM 1. Let E be a real vector space, $f : E \to \overline{\mathbb{R}}$ a function such that $C := \{x \in E \mid f(x) < +\infty\} \neq \emptyset$. Let the following assumptions hold:

i) $f(0) \geq 0$;

ii) there exist real numbers $\alpha > 0, \beta > 0$ such that
$f(\alpha x + \beta y) \leq \alpha f(x) + \beta f(y)$ $\forall x,y \in C$;

iii) $\limsup_{\lambda < 1, \lambda \to 1} f(\lambda x + (1-\lambda)y) \leq f(x)$ $\forall x,y \in C$.

Then there exists a function $\varphi : E \to \overline{\mathbb{R}}$, φ convex and concave with $\varphi(0) = 0$, such that $\varphi(x) \leq f(x)$ $\forall x \in E$. If we assume, in addition, that

iv) $C \subset \text{cone conv} (-C)$,

then there exists a linear function $\ell : E \to \mathbb{R}$ such that $\ell(x) \le f(x) \quad \forall x \in E$.

Proof. 1. We partially order the functions $g : E \to \overline{\mathbb{R}}$ by $g_1 \le g_2 :\Longleftrightarrow g_1(x) \le g_2(x) \quad \forall x \in E$, and consider the set

$$F := \{g : E \to \overline{\mathbb{R}} \mid g \text{ concave and positively homogeneous, } g \le f\}.$$

The set F is non-empty, since by assumption i) it contains the function g given by: $g(x) = -\infty$ for $x \neq 0$, $g(0) = 0$. Furthermore for any totally ordered subset $T \subset F$ there exists $\tilde{g} \in F$ with $\tilde{g} \ge g \quad \forall g \in T$ (simply choose $\tilde{g}(x) := \sup_{g \in T} g(x)$). Hence by Zorn's lemma F contains a maximal element φ, i.e., $\varphi \in F$ such that from $g \in F$ and $g \ge \varphi$ follows $g = \varphi$. We shall show that a maximal $\varphi \in F$ must be convex. Since $\varphi(0) = 0$ follows from positive homogeneity, this φ will then have all the properties to meet the conclusion of the first part of the theorem.

2. Let $\Lambda := \{(s,t) \in \mathbb{R}^2 \mid s \ge 0,\ t \ge 0, s + t = 1\}$ and $K := \alpha + \beta$. We shall derive from assumption ii) the existence of a dense subset $\Omega \subset \Lambda$ with the property that for all $(s,t) \in \Omega$ there exists an integer $n \ge 0$ such that the following statement, which we abbreviate by $\Psi(n,(s,t))$, holds true:

$$(1) \quad f(K^n(sx+ty)) \le K^n(sf(x)+tf(y)) \quad \forall x,y \in C.$$

We observe that $\Psi(0,(1,0))$ and $\Psi(0,(0,1))$ hold trivially. From assumption ii) follows that $K^{-1}f(Kz) \le f(z) \quad \forall z \in E$, and by induction $K^{-N}f(K^N z) \le K^{-n}f(K^n z)$ $\forall z \in E$ if N and n are integers with $N \ge n \ge 0$. Consequently $\Psi(n,(s,t))$ implies $\Psi(N,(s,t))$ for all $N \ge n$. Suppose now that $\Psi(n_i,(s_i,t_i))$ $(i=1,2)$ holds true. Set $n := \max\{n_1,n_2\}$ and

$$(s,t) := \alpha K^{-1}(s_1,t_1) + \beta K^{-1}(s_2,t_2) \in \Lambda.$$

Then it turns out that $\Psi(n+1,(s,t))$ is valid, too. Indeed, from assumption ii) and $\Psi(n,(s_i,t_i))$ we obtain

$$\begin{aligned} f(K^{n+1}(sx+ty)) &= f(\alpha K^n(s_1x+t_1y) + \beta K^n(s_2x+t_2y)) \\ &\le \alpha f(K^n(s_1x+t_1y)) + \beta f(K^n(s_2x+t_2y)) \\ &\le K^n\alpha(s_1f(x)+t_1f(y)) + K^n\beta(s_2f(x)+t_2f(y)) \\ &= K^{n+1}(sf(x)+tf(y)). \end{aligned}$$

Since (s,t) is a convex combination of (s_1,t_1) and (s_2,t_2) with coefficients depending only on α and β, we obtain in this way a dense subset Ω of Λ with the indicated property.

3. Let $\varphi \in F$. We show for all $s \ge 0, t \ge 0$ that

(2) $\varphi(sx+ty) \le sf(x) + tf(y) \quad \forall x,y \in C.$

If $(s,t) \in \Omega$, then (2) follows from (1) since $\varphi \le f$ and φ is positively homogeneous. If $(s,t) \in \Lambda \setminus \Omega$, then for $x,y \in C$ set $z := sx+ty$, and for $\lambda \in [0,1]$ let

$$x_\lambda := \frac{\lambda sx + (1-\lambda)z}{\lambda s + 1-\lambda}, \qquad (s_\lambda, t_\lambda) := \lambda(s,t) + (1-\lambda)(1,0).$$

Then $z = s_\lambda x_\lambda + t_\lambda y$, $(s_\lambda, t_\lambda) \in \Lambda$, and accordingly we have

$$\varphi(z) \le s_\lambda f(x_\lambda) + t_\lambda f(y) \text{ if } x_\lambda \in C,\ (s_\lambda, t_\lambda) \in \Omega.$$

Letting $\lambda \to 1$ we obtain in the limit, since Ω is dense in Λ and since $\limsup_{\lambda \uparrow 1} f(x_\lambda) \le f(x)$ by assumption iii), that $\varphi(z) \le sf(x) + tf(y)$. Hence (2) holds for all $(s,t) \in \Lambda$. But then (2) holds for all $s \ge 0, t \ge 0$, since both sides are positively homogeneous in (s,t).

4. Let $\varphi \in F$ be maximal. We show that φ is convex. Since φ is already concave and positively homogeneous the convexity of φ follows if

$$\varphi(z) \ge -\varphi(-z) \quad \forall z \in E.$$

Assume, for contradiction, that $\varphi \in F$ is maximal, but not convex. Accordingly then there exists $z_o \in E$ and $c \in \mathbb{R}$ such that

(3) $\varphi(z_o) < c < -\varphi(-z_o).$

Define for all $x \in E$

$$g(x) := \sup \{\varphi(\xi) + tc \mid x = \xi + tz_o,\ \xi \in E,\ t \ge 0\}.$$

Then $g : E \to \overline{\mathbb{R}}$ is concave and positively homogeneous. Indeed, if $t_i \ge 0$ and $\varphi(x_i - t_i z_o) > -\infty$ $(i=1,2)$ we obtain

$$\begin{aligned}\varphi(x_1 - t_1 z_o) + t_1 c + \varphi(x_2 - t_2 z_o) + t_2 c &\le \varphi(x_1 + x_2 - (t_1+t_2)z_o) + (t_1+t_2)c \\ &\le g(x_1 + x_2),\end{aligned}$$

and therefore $g(x_1) + g(x_2) \le g(x_1+x_2)$ if $g(x_1) > -\infty$, $g(x_2) > -\infty$. Likewise we verify for any $\lambda > 0$ that $g(\lambda x) = \lambda g(x) \ \forall x \in E$. And

$$g(0) = \sup_{t \ge 0} t(\varphi(-z_o) + c) = 0,$$

since $\varphi(-z_o) + c \le 0$ by the second inequality in (3). Furthermore we have $g(x) \ge \varphi(x) \ \forall x \in E$ (fix $t = 0$ in the definition of g) and $g(z_o) \ge c > \varphi(z_o)$ (fix $t = 1$, $\xi = 0$ in the definition of g and use the first inequality in (3)). Hence $g \ge \varphi$, but $g \ne \varphi$. Would we have $f \ge g$, then this would contradict the maximality of φ in F. Hence $f(x) < g(x)$ for some $x \in E$, i.e., there exists $x \in C$ and $t > 0$ with

(4) $f(x) < \varphi(x - tz_o) + tc.$

If we replace in the definition of g the pair (z_o,c) by $(-z_o,-c)$ and accordingly define for all $y \in E$

$$g(y) := \sup \{\varphi(\eta) - sc \mid y = \eta - sz_o,\ \eta \in E,\ s \geq 0\},$$

then again g is concave and positively homogeneous, with $g \geq \varphi$ and $g(-z_o) > \varphi(-z_o)$, and an analogous argument as before gives $y \in C$ and $s > 0$ with

$$(5) \quad f(y) < \varphi(y+sz_o) - sc.$$

From (4) and (5) follows

$$(6) \quad sf(x) + tf(y) < s\varphi(x-tz_o) + t\varphi(y+sz_o).$$

But using $sx + ty = s(x-tz_o) + t(y+sz_o)$ we obtain from (2)

$$\begin{aligned} sf(x) + tf(y) &\geq \varphi(sx+ty) \\ &= \varphi(s(x-tz_o) + t(y+sz_o)) \\ &\geq s\varphi(x-tz_o) + t\varphi(y+sz_o), \end{aligned}$$

since φ is concave and positively homogeneous, and since $\varphi(x-tz_o) > -\infty$, $\varphi(y+sz_o) > -\infty$ by (4) and (5). This contradicts (6). Thus, if $\varphi \in F$ is maximal, then φ must also be convex. This proves the first part of the theorem.

5. Now we exploit assumption iv). Take the function φ given by the first part of the theorem. It satisfies in particular $\varphi(\lambda x) = \lambda\varphi(x)$ $\forall \lambda \in \mathbb{R}$, $\forall x \in E$. Set $L := \{x \in E \mid \varphi(x) \in \mathbb{R}\}$. L is a linear subspace, and φ is linear on L. From $\varphi \leq f$ follows $\varphi < +\infty$ on cone conv C and $\varphi > -\infty$ on cone conv (-C). Hence

$$\text{cone conv } C \cap \text{cone conv } (-C) \subset L.$$

From assumption iv) follows

$$C \subset \text{cone conv } C \cap \text{cone conv } (-C).$$

Hence $C \subset L$. Since $\varphi|_L$ is linear, and since f equals $+\infty$ outside C, we can trivially extend $\varphi|_L$ to a linear function $\ell : E \to \mathbb{R}$ with $\ell \leq f$. The second part of the theorem is also proved. □

The possibility $f(0) = +\infty$ is permitted by assumptions i) - iii). However, the addition of assumption iv) has the side-effect that $f(0) < +\infty$. Indeed: From (1) follows for all $(s,t) \in \Omega$ that there exists $\rho > 0$ such that $x,y \in C \Rightarrow \rho(sx+ty) \in C$. By assumption iii) this holds even for all $s \geq 0$, $t \geq 0$ with $(s,t) \neq 0$. Assumption iv) implies then, since $C \neq \emptyset$, that $0 = \Sigma\, s_i x_i$ with $x_i \in C$, $s_i \geq 0$ and not all zero. Hence $0 \in C$.

Let us note that assumption iii) is automatically satisfied, if f is a convex function.

III. A SANDWICH THEOREM

We derive now from Theorem 1 a sandwich version. For the convex case this result has been proved by Borwein [6].

THEOREM 2. Let E be a real vector space; let $f : E \to \overline{\mathbb{R}}$ and $g : E \to \overline{\mathbb{R}}$ be functions such that $C := \{x \in E \mid f(x) < +\infty\} \neq \emptyset$ and $D := \{x \in E \mid g(x) > -\infty\} \neq \emptyset$. Let the following assumptions hold:

i) $g(x) \leq f(x) \quad \forall x \in E$;

ii) there exist real numbers $\alpha > 0$, $\beta > 0$ such that for all $x,y \in C$ and $u,v \in D$

$$\inf \{f(z) - g(w) \mid (z,w) \in C \times D,\ z-w = \alpha(x-u) + \beta(y-v)\}$$
$$\leq \alpha(f(x) - g(u)) + \beta(f(y) - g(v));$$

iii) for all $x,y \in C$ and $u,v \in D$

$$\limsup_{\lambda<1,\lambda\to1} f(\lambda x+(1-\lambda)y) \leq f(x), \quad \liminf_{\lambda<1,\lambda\to1} g(\lambda u+(1-\lambda)v) \geq g(u).$$

Then there exists a function $\psi : E \to \overline{\mathbb{R}}$, ψ convex and concave, such that

$$g(x) \leq \psi(x) \leq f(x) \quad \forall x \in E.$$

Proof. We apply Theorem 1 to the function $p : E \to \overline{\mathbb{R}}$ defined by

$$p(\xi) := \inf \{f(z) - g(w) \mid (z,w) \in C \times D,\ z-w = \xi\}.$$

Clearly $p(\xi) < +\infty \Longleftrightarrow \xi \in C - D$. Assumption i) gives $p(0) \geq 0$. Let $\xi,\eta \in C - D$. From assumption ii) follows for any $x,y \in C$ and $u,v \in D$ satisfying $\xi = x-u$, $\eta = y-v$ that $p(\alpha\xi+\beta\eta) \leq \alpha(f(x) - g(u)) + \beta(f(y) - g(v))$, and taking on the right-hand side of this inequality the infimum with regard to all such x,u and y,v we obtain

$$p(\alpha\xi+\beta\eta) \leq \alpha p(\xi) + \beta p(\eta) \quad \forall \xi,\eta \in C - D.$$

Likewise it follows from assumption iii) that

$$\limsup_{\lambda<1,\lambda\to1} p(\lambda\xi+(1-\lambda)\eta) \leq p(\xi) \quad \forall \xi,\eta \in C - D.$$

Hence we may apply Theorem 1 to the function p, and we obtain a function $\varphi : E \to \overline{\mathbb{R}}$ such that $\varphi(\xi) \leq p(\xi) \ \forall \xi \in E$, which in view of the definition of p gives

$$\varphi(x-u) \leq f(x) - g(u) \quad \forall x \in C, \quad \forall u \in D.$$

The function φ is convex and concave and satisfies $\varphi(0) = 0$, whence for any $a \in E$ we have that $\varphi(x-a) - \varphi(u-a) \leq \varphi(x-u)$, provided $\varphi(x-a) > -\infty$, $\varphi(u-a) < +\infty$. Therefore we obtain for any $a \in E$

$$(7) \quad g(u) - \varphi(u-a) \leq f(x) - \varphi(x-a) \quad \forall x \in C \text{ with } \varphi(x-a) > -\infty \text{ and } \forall u \in D \text{ with } \varphi(u-a) < +\infty.$$

Let $a \in E$ be such that for both sides or for neither side of the inequality (7) there exist feasible points x,u. Then there exists $c \in \mathbb{R}$ such that $g(u) - \varphi(u-a) \leq c \leq f(x) - \varphi(x-a)$ for the same x,u as in (7). Setting now $\psi(\xi) := \varphi(\xi-a)+c$ we see that $g \leq \psi \leq f$. Hence ψ meets the conclusion of the theorem. It remains to settle the case where for each $a \in E$ exactly one side of inequality (7) admits feasible points. Then the two sets

$$\tilde{C} := \{a \in E \mid \exists\ x \in C \text{ with } \varphi(x-a) > -\infty\}, \quad \tilde{D} := \{a \in E \mid \exists\ u \in D \text{ with } \varphi(u-a) < +\infty\}$$

are complementary, i.e., $\tilde{C} \cap \tilde{D} = \emptyset$, $E = \tilde{C} \cup \tilde{D}$. From $\varphi(0) = 0$ follows $C \subset \tilde{C}$ and $D \subset \tilde{D}$. Define $\psi(a) := -\infty$ if $a \in \tilde{C}$, $\psi(a) := +\infty$ if $a \in \tilde{D}$. Then trivially $g \leq \psi \leq f$. $\tilde{C}$ and $\tilde{D}$ have convex complements and are therefore themselves convex. For example the complement of $\tilde{C}$ equals $\bigcap_{x \in C} \{a \in E \mid \varphi(x-a) = -\infty\}$ which as an intersection of convex sets is convex. Hence ψ so defined is simultaneously convex and concave. □

We obtain now from Theorem 2 a generalization of Stone's lemma [5]. Recall that two subsets A,B in a vector space E are said to be complementary if $A \cup B = E$ and $A \cap B = \emptyset$.

COROLLARY 1. Let Γ, Δ be nonempty subsets of a real vector space E with the following properties:

i) $\Gamma \cap \Delta = \emptyset$;

ii) there exist real numbers $\alpha > 0$, $\beta > 0$ such that $\alpha(\Gamma-\Delta) + \beta(\Gamma-\Delta) \subset \Gamma-\Delta$;

iii) for all $\xi, \eta \in \Gamma-\Delta$ is $[\xi,\eta] \cap (\Gamma-\Delta)$ open in $[\xi,\eta]$.

Then there exist complementary convex subsets $\tilde{\Gamma}, \tilde{\Delta}$ of E such that $\Gamma \subset \tilde{\Gamma}$, $\Delta \subset \tilde{\Delta}$.

Proof. We apply Theorem 2 (with an obvious modification in assumption iii)) by setting

$f = -\infty$ on Γ, $f = +\infty$ outside Γ,

$g = +\infty$ on Δ, $g = -\infty$ outside Δ,

and with the resulting function ψ such that $g \leq \psi \leq f$ we define

$\tilde{\Gamma} := \{x \in E \mid \psi(x) \leq 0\}$, $\tilde{\Delta} := \{x \in E \mid \psi(x) > 0\}$. $\tilde{\Gamma}$ and $\tilde{\Delta}$ have the required properties. □

Theorem 2 can in turn be derived from the Corollary 1 if we add the assumption $C \cap D \neq \emptyset$. Indeed, with the data of Theorem 2 set $\Gamma := \{(x,r) \in E \times \mathbb{R} \mid f(x) < r\}$, $\Delta := \{(x,r) \in E \times \mathbb{R} \mid g(x) > r\}$. Then by Corollary 1 there exist complementary convex sets $\tilde{\Gamma}$ and $\tilde{\Delta}$ such that $\Gamma \subset \tilde{\Gamma}, \Delta \subset \tilde{\Delta}$. One verifies easily that $C \cap D \neq \emptyset$ implies for all $x \in E$ that $(x,r_1) \in \tilde{\Gamma}$ and $(x,r_2) \in \tilde{\Delta}$ is only possible if $r_1 > r_2$. Hence one has for all $x \in E$ that

$$\inf \{r \mid (x,r) \in \tilde{\Gamma}\} = \sup \{r \mid (x,r) \in \tilde{\Delta}\} =: \psi(x),$$

and ψ satisfies the conclusion of Theorem 2.

IV. OPTIMIZATION PROBLEMS

We apply now Theorem 1 to a general class of optimization problems and obtain a multiplier result of Kuhn-Tucker type. Recall that lin A (the linear hull of A) denotes the smallest linear subspace containing the set A.

THEOREM 3. Let $X \neq \emptyset$ be an arbitrary set, E a real vector space, $f : X \to \overline{\mathbb{R}}$ a function, $G : X \to 2^E$ a set-valued mapping such that $A := \{x \in X \mid G(x) \neq \emptyset$ and $f(x) < +\infty\} \neq \emptyset$. Let the following assumptions hold:

i) $0 \leq \inf \{f(x) \mid x \in X,\ 0_E \in G(x)\}$;

ii) there exist real numbers $\alpha > 0$, $\beta > 0$ such that for all $x,y \in A$ and all $\xi \in G(x)$, $\eta \in G(y)$ holds $\inf \{f(z) \mid z \in X,\ \alpha\xi + \beta\eta \in G(z)\} \leq \alpha f(x) + \beta f(y)$;

iii) for all $x,y \in A$, $\xi \in G(x)$, $\eta \in G(y)$ holds
$$\limsup_{\lambda<1,\lambda\to 1} \inf \{f(z) \mid z \in X,\ \lambda\xi + (1-\lambda)\eta \in G(z)\} \leq f(x).$$

Then there exists a function $\varphi : E \to \overline{\mathbb{R}}$, φ convex and concave with $\varphi(0) = 0$, such that $\varphi(\xi) \leq f(x)\ \forall \xi \in G(x),\ \forall x \in X$. The function φ is real-valued and linear on the subspace $L := \text{lin}\ (\text{cone conv}\ G(A) \cap \text{cone conv}\ -G(A))$.

Proof. We apply Theorem 1 to the function $p : E \to \overline{\mathbb{R}}$ given by

$$p(\xi) := \inf \{f(x) \mid x \in X,\ \xi \in G(x)\}.$$

Clearly $p(\xi) < +\infty \Leftrightarrow \xi \in G(A)$. Assumption i) gives $p(0) \geq 0$. Assumption ii) gives $p(\alpha\xi+\beta\eta) \leq \alpha f(x) + \beta f(y)$, and taking on the right-hand side of this inequality the infimum with regard to all x,y such that $\xi \in G(x)$, $\eta \in G(y)$ we obtain

$$p(\alpha\xi+\beta\eta) \leq \alpha p(\xi) + \beta p(\eta) \quad \forall \xi,\eta \in G(A).$$

Likewise assumption iii) gives $\limsup_{\lambda\uparrow 1} p(\lambda\xi+(1-\lambda)\eta) \leq p(\xi) \quad \forall \xi \in G(A)$. From Theorem 1 we obtain the function $\varphi : E \to \overline{\mathbb{R}}$ with $\varphi(\xi) \leq p(\xi) \quad \forall \xi \in E$, which in view of the definition of p gives

$$\varphi(\xi) \leq f(x) \quad \forall \xi \in G(x), \quad \forall x \in X.$$

That φ is real-valued and hence linear on the subspace L follows easily as in the last paragraph of the proof of Theorem 1. □

We remark that the conclusion of Theorem 3 is in turn a sufficient condition for the validity of its statement i). In the following corollary we specialize the results of Theorem 3 to the case of ordinary programming problems with a finite number of inequality constraints. In dealing with $\overline{\mathbb{R}}^n$ we use the natural (componentwise) ordering.

COROLLARY 2. Let $X \neq \emptyset$ be an arbitrary set. Let $f : X \to \overline{\mathbb{R}}$ and $t_i : X \to \mathbb{R} \cup \{-\infty\}$ $(i=1,\ldots,n)$ be functions. Let $A := \{x \in X \mid f(x) < +\infty\} \neq \emptyset$ and and $t(x) := (t_1(x),\ldots,t_n(x))$.

i) $0 \leq \inf \{f(x) \mid x \in X,\ t(x) \leq 0\}$;

We make the following assumptions:

ii) there exist real numbers $\alpha > 0$, $\beta > 0$ such that for all $x,y \in A$ holds $\inf \{f(z) \mid z \in X,\ t(z) \leq \alpha t(x) + \beta t(y)\} \leq \alpha f(x) + \beta f(y)$;

iii) for all $x,y \in A$ holds $\limsup_{\lambda<1,\lambda\to 1} \inf \{f(z) \mid z \in X,\ t(z) \leq \lambda t(x) + (1-\lambda)t(y)\} \leq f(x)$;

iv) there exists $x^o \in A$ with $t(x^o) < 0$.

Then there exist real numbers $c_i \geq 0$ $(i=1,\ldots,n)$ such that

$$-\textstyle\sum_i c_i t_i(x) \leq f(x) \quad \forall x \in X.$$

Proof. We reduce it to Theorem 3, with $E = \mathbb{R}^n$, defining the set-valued mapping $G(\cdot)$ by

$$\xi \in G(x) :\Longleftrightarrow \xi \geq t(x).$$

Then $G(A) = t(A) + \mathbb{R}^n_+$. Assumptions i) - iii) ensure that the corresponding assumptions of Theorem 3 are met. Assumption iv) implies that $0 \in \operatorname{int} G(A)$, and hence cone conv $G(A) = \mathbb{R}^n$. Accordingly the function φ furnished by Theorem 3 is linear on all of $\mathbb{R}^n$. Let φ be represented as $\varphi(\xi) = -\sum_i c_i\xi_i$. Then we have

$$-\sum_i c_i\xi_i \le f(x) \quad \forall x \in X, \quad \forall \xi \in \mathbb{R}^n \text{ such that } \xi \ge t(x).$$

This implies $c_i \ge 0 \;\; \forall i$, since $f \not\equiv +\infty$. The desired conclusion follows at once. □

It should be noted that Corollary 2 is no longer true if we admit for $t_i(x)$ the value $+\infty$ instead of $-\infty$.

We conclude with a result, which is already contained in [1]; it gives a nice example where assumptions iii) and iv) in Corollary 2 are automatically fulfilled.

COROLLARY 3. Let $X \neq \emptyset$ be an arbitrary set, and $f_1,\ldots,f_n : X \to \mathbb{R} \cup \{-\infty\}$. Let there exist $\alpha > 0$, $\beta > 0$ such that for all $x,y \in X$ and $\varepsilon > 0$ there exists $z \in X$ with

$$f_i(z) \le \alpha f_i(x) + \beta f_i(y) + \varepsilon \quad \forall i=1,\ldots,n.$$

Let $\max(f_1,\ldots,f_n) \ge 0$ on X. Then there exist real numbers $c_1 \ge 0,\ldots,c_n \ge 0$ with $c_1 + \ldots + c_n = 1$ such that $c_1 f_1 + \ldots + c_n f_n \ge 0$ on X.

Proof. With $h \in \mathbb{R}$ as an additional variable the condition that $\max(f_1,\ldots,f_n) \ge 0$ on X can be written as

$$0 \le \inf\, \{h \mid (x,h) \in X \times \mathbb{R},\ f_i(x) - h \le 0 \ (i=1,\ldots,n)\}.$$

Set $f(x,h) := h$, $t_i(x,h) := f_i(x) - h$ and apply Corollary 2. From it we obtain real numbers $c_1 \ge 0,\ldots,c_n \ge 0$ such that

$$-\sum_i c_i(f_i(x) - h) \le h \quad \forall x \in X, \quad \forall h \in \mathbb{R}.$$

Variation of h alone shows that $\sum c_i = 1$, and it remains

$$-\sum_i c_i f_i(x) \le 0 \quad \forall x \in X. \quad \square$$

REFERENCES

1. B. Fuchssteiner, H. König, New versions of the Hahn-Banach theorem. In: "General Inequalities 2", edited by E.F. Beckenbach, pp. 255-266. Birkhäuser, Basel, 1980.

2. H. König, On some basic theorems in convex analysis. In: "Modern Applied Mathematics", edited by B. Korte. North-Holland, Amsterdam, 1982.

3. W. Oettli, On a new version of the Hahn-Banach theorem. To appear in: Proceedings of the International Congress on Mathematical Programming, Rio de Janeiro, 1981. North-Holland, Amsterdam.

4. W. Oettli, Optimality conditions for programming problems involving multivalued mappings. In: "Modern Applied Mathematics", edited by B. Korte. North-Holland, Amsterdam, 1982.

5. R.B. Holmes, Geometric Functional Analysis and its Applications, p. 7. Springer, New York, 1975.

6. J.M. Borwein, On the Hahn-Banach Sandwich Principle. Research Report 81-2, Department of Mathematics, Carnegie-Mellon University, Pittsburgh, 1981.

A NOTE ON THE CHEBYSHEV ε-APPROXIMATION PROBLEM

J.-J. Strodiot, V. Hien Nguyen, N. Heukemes

Facultés Universitaires de Namur
Namur, Belgium

I. INTRODUCTION

This paper is concerned with the approximation of continuous functions in the Chebyshev norm. We let K be a compact topological space and we denote by C(K) the space of all real-valued continuous functions defined on K. We let $f, f_1, \ldots, f_n$ be linearly independent functions in C(K) and we denote by V the linear subspace spanned by $f_1, \ldots, f_n$. Let $\varepsilon \geq 0$. With Buck [2] and Singer [6], the following ε-approximation **problem** is considered : Find an element $\bar{g}$ of V such that

$$\|f-\bar{g}\|_\infty \leq \|f-g\|_\infty + \varepsilon \text{ for all } g \in V.$$

Here $\|.\|_\infty$ denotes the Chebyshev norm defined for each $h \in C(K)$ by

$$\|h\|_\infty = \max_{k \in K} |h(k)|.$$

Such an element $\bar{g}$ is called an ε-best approximation to f from V. It is highly desirable to have some characterization property which allows us to recognize an ε-best approximation.

When $\varepsilon = 0$ numerous characterizations of best approximations have been obtained, especially in the case where K is a compact interval of $\mathbb{R}$, see [4] and the references quoted therein. One of the most important results is the Chebyshev alternation theorem which allows to derive the Remes algorithm.

One way to arrive at these characterizations is to observe that $\bar{g} = \sum_{j=1}^{n} \bar{x}_j f_j$ is a best approximation to f from V ($\varepsilon = 0$) if and only if $\bar{x} = (\bar{x}_1, \ldots, \bar{x}_n)$ is a solution to the problem of minimizing

$$\mu(x) = \|f - \sum_{j=1}^{n} x_j f_j\|_\infty = \max_{\substack{k \in K \\ r = \pm 1}} r(f(k) - \sum_{j=1}^{n} x_j f_j(k))$$

over all $x = (x_1, \ldots, x_n) \in \mathbb{R}^n$, hence if and only if 0 belongs to $\partial\mu(\bar{x})$, the subdifferential of μ at $\bar{x}$ [5]. Thus, for characterizing $\bar{x}$ one needs to calculate the subdifferential of a "max function" by using the Rockafellar-Valadier formula (see, for example, [8, Theorem 2]).

In this paper our aim is to present a Chebyshev alternation theorem for an ε-best approximation with $\varepsilon > 0$. More precisely, we want to show step-by-step what conditions are necessary for deriving the well-known alternation properties. This time, for characterizing $\bar{x}$, we shall need to calculate the ε-subdifferential of the "max function" μ. Therefore, we first begin by recalling this formula [7].

Let T be a compact topological space, and let F(t,x) be a real-valued function defined on $T \times \mathbb{R}^n$ convex (hence continuous) in x for every $t \in T$ and upper semicontinuous in t for every $x \in \mathbb{R}^n$. We set $F(x) = \max_{t \in T} F_t(x)$, where $F_t(.)$ denotes the function on $\mathbb{R}^n$ defined by the equality $F_t(x) = F(t,x)$. Let $\bar{x} \in \mathbb{R}^n$. Then $x^* \in \partial_\varepsilon F(\bar{x})$ if and only if there exist n+1 points $t_1, \ldots, t_{n+1} \in T$, scalars $\lambda_i \geq 0$ for $i = 1, \ldots, n+1$, with $\sum_{i=1}^{n+1} \lambda_i = 1$, and scalars $\varepsilon_i \geq 0$ for $i = 1, \ldots, n+1$ such that :

$$\text{(i)} \quad x^* \in \sum_{i=1}^{n+1} \partial_{\varepsilon_i} (\lambda_i F_{t_i})(\bar{x}) ,$$

$$\text{(ii)} \quad 0 \leq F(\bar{x}) - \sum_{i=1}^{n+1} \lambda_i F_{t_i}(\bar{x}) \leq \varepsilon - \sum_{i=1}^{n+1} \varepsilon_i .$$

Finally, following Dem'yanov and Malozemov [3], we introduce the approximate subdifferential $L_\varepsilon(\bar{x})$ of the function F at $\bar{x}$ as the convex hull of the set

$$\bigcup_{t \in T_\varepsilon(\bar{x})} \partial F_t(\bar{x})$$

where $T_\varepsilon(\bar{x}) = \{t \in T \mid F(\bar{x}) - F_t(\bar{x}) \leq \varepsilon\}$. Then, we have $L_\varepsilon(\bar{x}) \subset \partial_\varepsilon f(\bar{x})$.

II. APPLICATION TO THE APPROXIMATION THEORY

In this section characterizations of ε-best approximations to f from V are derived from the formula of the ε-subdifferential of a "max function".

PROPOSITION 1. Let $\bar{x} = (\bar{x}_1, \ldots, \bar{x}_n)^T \in \mathbb{R}^n$. The function $\bar{g} = \sum_{j=1}^{n} \bar{x}_j f_j$ is an ε-best approximation to f from V if and only if there exist n+1 points $k_1, \ldots, k_{n+1}$ of K, scalars $r_1, \ldots, r_{n+1} \in \{-1, 1\}$, and $\lambda_i \geq 0$ for $i = 1, \ldots, n+1$, with $\sum_{i=1}^{n+1} \lambda_i = 1$, such that

$$\text{(i) } 0 = \sum_{i=1}^{n+1} \lambda_i r_i f_j(k_i) \text{ for } j = 1, \ldots, n$$

$$\text{(ii) } \mu(\bar{x}) - \varepsilon \leq \sum_{i=1}^{n+1} \lambda_i r_i e(k_i) \leq \mu(\bar{x}),$$

where $e(k_i)$ denotes the error $f(k_i) - \sum_{j=1}^{n} \bar{x}_j f_j(k_i)$ for $i = 1, \ldots, n+1$.

Proof. Clearly $\bar{g} = \sum_{j=1}^{n} \bar{x}_j f_j$ is an ε-best approximation to f from V if and only if $\bar{x} = (\bar{x}_1, \ldots, \bar{x}_n)^T \in \mathbb{R}^n$ is an ε-solution to the problem of minimizing $\mu(x)$ over all $x \in \mathbb{R}^n$, hence if and only if $0 \in \partial_\varepsilon \mu(\bar{x})$. But then, using the analogue of the Rockafellar - Valadier formula recalled in Introduction, this condition holds true if and only if there exist n+1 points $k_1, \ldots, k_{n+1}$ of K, scalars $r_1, \ldots, r_{n+1} \in \{-1, 1\}$, $\lambda_i \geq 0$ for $i = 1, \ldots, n+1$, with $\sum_{i=1}^{n+1} \lambda_i = 1$, and $\varepsilon_i \geq 0$ for $i = 1, \ldots, n+1$ such that

$$(1) \qquad 0 \in \sum_{i=1}^{n+1} \partial_{\varepsilon_i} (\lambda_i r_i e(k_i))(\bar{x}),$$

$$(2) \qquad \mu(\bar{x}) - \varepsilon + \sum_{i=1}^{n+1} \varepsilon_i \leq \sum_{i=1}^{n+1} \lambda_i r_i e(k_i) \leq \mu(x).$$

Since the $e(k_i)$'s are linear in x, we have for i = 1,..., n+1,

$$\partial_{\varepsilon_i}(\lambda_i\, r_i\, e(k_i))(\bar{x}) = -\lambda_i\, r_i (f_1(k_i),\ldots, f_n(k_i))^T.$$

Substituting these expressions into (1), whe have condition (i). Condition (ii) follows immediately from (2) because $\varepsilon_i \geq 0$. So the first part of the proof has been proved.
Conversely if conditions (i) and (ii) are assumed to be satisfied, then (1) and (2) hold true with ε_i for i = 1,..., n+1, and, following the remark made at the beginning, the proof is finished. □

Now if the system $\{f_1,\ldots, f_n\}$ satisfies the *Haar condition* on K, i.e., for any n distinct points $k_1,\ldots, k_n$ of K the vectors $(f_1(k_i),\ldots, f_n(k_i))^T$ for i = 1,..., n are linearly independent in $\mathbb{R}^n$ - see, for example, Laurent [4]-, then a strengthened version of Proposition 1 can be obtained.

PROPOSITION 2. Assume that the system $\{f_1,\ldots, f_n\}$ satisfies the Haar condition. If $\varepsilon < \mu(\bar{x})$, then in Proposition 1 the points $k_1,\ldots, k_{n+1}$ of K are distinct, and $\lambda_i > 0$ for i = 1,..., n+1.

Proof. By contradiction, suppose that the n+1 points $k_1,\ldots, k_{n+1}$ of K are not distinct. Then, either $k_1 = \ldots = k_{n+1}$ or q points, among the k_i's, are distinct. To simplify the notation, suppose that these points are $k_1,\ldots, k_q$, with $1 < q < n+1$. Note that the first situation corresponds to the case q = 1. Let us introduce the following notation for ℓ = 1,..., q,

$$N_\ell = \{i \mid i = 1,\ldots, n+1 \text{ such that } k_i = k_\ell\},$$

$$\bar{\lambda}_\ell = \sum_{i \in N_\ell} \lambda_i \quad , \quad \text{and}$$

$$\bar{r}_\ell = \begin{cases} (1/\bar{\lambda}_\ell) \sum_{i \in N_\ell} \lambda_i\, r_i & \text{if } \bar{\lambda}_\ell \neq 0, \\ 0 & \text{if } \bar{\lambda}_\ell = 0. \end{cases}$$

With these definitions conditions (i) and (ii) of Proposition 1 can now be expressed as

(3) $$0 = \sum_{\ell=1}^{q} \bar{\lambda}_\ell \, \bar{r}_\ell \, f_j(k_\ell) \text{ for } j = 1,\ldots, n,$$

(4) $$\mu(\bar{x}) - \varepsilon \le \sum_{\ell=1}^{q} \bar{\lambda}_\ell \, \bar{r}_\ell \, e(k_\ell) \quad \mu(\bar{x}),$$

respectively, where we have gathered together the λ_i's and r_i's corresponding to the identical points. Observe that there is at least one ℓ for which $\bar{r}_\ell \ne 0$. To see this, let $\bar{r}_\ell = 0$ for all ℓ ; then, by (4), we have $\mu(\bar{x}) \le \varepsilon$, contradicting our assumption. Therefore $\bar{r}_\ell \ne 0$ for at least one ℓ. In fact, we claim that $\bar{r}_\ell \ne 0$ for all $\ell = 1,\ldots, q$. Indeed, if not, then without loss of generality we may suppose that $\bar{r}_\ell \ne 0$ for $\ell = 1,\ldots, s$, and $\bar{r}_\ell = 0$ for $\ell = s+1,\ldots, q$, with $s < q$. From (3) it follows that

(5) $$\sum_{\ell=1}^{s} \bar{\lambda}_\ell \, \bar{r}_\ell \, f_j(k_\ell) = 0 \text{ for } j = 1,\ldots, n.$$

Since the vectors $\bar{r}_\ell(f_\ell(k_\ell),\ldots, f_n(k_\ell))^T$ for $\ell = 1,\ldots, s$ are linearly independent because the Haar condition holds true, we deduce from (5) that $\bar{\lambda}_\ell = 0$ for $\ell = 1,\ldots, s$. But then this implies, by (4), that $\mu(\bar{x}) \le \varepsilon$, since $\bar{r}_\ell = 0$ for $\ell = s+1,\ldots, q$. This contradicts the assumption that $\varepsilon < \mu(\bar{x})$. Thus $\bar{r}_\ell \ne 0$ for all $\ell = 1,\ldots,q$.
Consequently, by the Haar condition again, the vectors $\bar{r}_\ell(f_1(k_\ell),\ldots,f_n(k_\ell))^T$ for $\ell = 1,\ldots, q$ are linearly independent. But this fact contradicts (3) because we know that $\sum_{\ell=1}^{q} \bar{\lambda}_\ell = 1$. Therefore $q = n+1$, and the points $k_1,\ldots, k_{n+1}$ are distinct.
Finally, from the Haar condition it follows immediately that $\lambda_i > 0$ for $i = 1,\ldots, n+1$, and this completes the proof. □

Remark 1. It is interesting to note that if $\varepsilon \ge \mu(\bar{x})$ then the points $k_1,\ldots, k_{n+1}$ of K as appeared in Proposition 2 may be identical, even if the system $\{f_1,\ldots, f_n\}$ satisfies the Haar condition. For example, if $K = [0,1] \subset \mathbb{R}$, $f(x) = x$, and V is the one-dimensional linear subspace of $C[0,1]$ spanned by the constant function $f_1(x) = 1$, then V satisfies obviously the Haar condition. For $\varepsilon = 1$, f_1 is an ε-best approximation to f from V, $\bar{x} = 1$, and $1 = \mu(\bar{x}) \le \varepsilon$. The reader can easily verify that the identical points $k_1 = k_2 = 1$ in $[0,1]$ satisfy conditions (i) and (ii) of Proposition 1 with $r_1 = 1$, $r_2 = -1$, and $\lambda_1 = \lambda_2 = 1/2$.

The characterizations derived in this manner take on an especially useful form in the case that K is a compact interval [a, b] of real numbers. We state first the following.

COROLLARY 1. Let K be a compact interval [a, b] of real numbers. Let $\{f_1,\ldots, f_n\}$ have the Haar property on [a, b]. Assume $\varepsilon < \mu(\bar{x})$. Then, in Proposition 1, the scalars r_i for $i = 1,\ldots, n+1$ alternate in sign (in addition to the conclusion that the points $k_1,\ldots, k_{n+1}$ in [a, b] are distinct and that $\lambda_i > 0$ for $i = 1,\ldots, n+1$).

Proof. Since $\sum_{i=1}^{n+1} \lambda_i r_i f_j(k_i) = 0$ for $j = 1,\ldots, n$, it follows that $\sum_{i=1}^{n+1} \lambda_i r_i h(k_i) = 0$ for all $h \in V$. As $\lambda_i r_i \neq 0$ for $i = 1,\ldots, n+1$, we deduce from Laurent [4, Proposition 3.5.1] that the $\lambda_i r_i$'s are alternating positive and negative. Since the λ_i's are > 0, the r_i's alternate also in sign. This completes the proof. □

Although in Corollary 1 the r_i's are alternating positive and negative, we do not have that the errors $e(k_i)$ for $i = 1,\ldots, n+1$ alternate also in sign. For instance, if we set $\varepsilon = 1/2$ in the preceding example of Remark 1, then f_1 is still an ε-best approximation to f from V. However, the difference $f-f_1$ has always the same sign. Consequently, in order to obtain this alternation property, it is clear that we have to strengthen our assumptions. Following Dem'yanov and Malozemov [3], a point $\bar{x} \in \mathbb{R}^n$ is said to be an ε-stationary point of μ if $0 \in L_\varepsilon(\bar{x})$. Obviously, every ε-stationary point $\bar{x}$ of μ is such that $\bar{g} = \sum_{j=1}^{n} \bar{x}_j f_j$ is an ε-best approximation to f from V. But the converse is not true.

PROPOSITION 3 (ε-Alternation Theorem). Let $\bar{x} = (\bar{x}_1,\ldots, \bar{x}_n)^T \in \mathbb{R}^n$. Let K be a compact interval $[a, b] \subset \mathbb{R}$. Suppose that V satisfies the Haar condition, and that $\varepsilon < \mu(\bar{x})$. Then $\bar{x}$ is an ε-stationary point of μ if and only if there exist n+1 points $k_1,\ldots, k_{n+1}$ in [a,b], with $a \leq k_1 < k_2 < \ldots < k_n < k_{n+1} \leq b$, such that

(i) $e(k_i) = -e(k_{i+1})$ for $i = 1,\ldots, n$

(ii) $\mu(\bar{x}) - \varepsilon \leq |e(k_i)| \leq \mu(\bar{x})$ for $i = 1,\ldots, n+1$.

Proof. By the definition of $L_\varepsilon(\bar{x})$ - see Introduction, $\bar{x}$ is an ε-stationary point of μ if and only if there exist n+1 points $k_1,\ldots, k_{n+1}$ in [a,b], scalars $r_i \in \{-1,1\}$ and $\lambda_i \geq 0$ for i = 1,..., n+1, with $\sum_{i=1}^{n+1} \lambda_i = 1$, such that

$$(6) \qquad 0 \in \sum_{i=1}^{n+1} \partial(\lambda_i \, r_i \, e(k_i))(\bar{x}),$$

$$(7) \qquad \mu(\bar{x}) - \varepsilon \leq r_i \, e(k_i) \leq \mu(\bar{x}) \text{ for } i = 1,\ldots, n+1.$$

Since $\mu(\bar{x}) - \varepsilon > 0$, it follows from (7) that

$$(8) \qquad r_i = \text{sign of } e(k_i).$$

Thus (7) is equivalent to condition (ii) of the proposition. On the other hand, since V satisfies the Haar condition and since $\varepsilon < \mu(\bar{x})$, by a reasoning similar to that of Proposition 2, it can be shown that $k_1 < k_2 < \ldots < k_{n+1}$, and that the r_i's are alternating positive and negative. Consequently, condition (i) of the proposition follows from (6) and (7), because of (8). What remains is to prove that (6) is a consequence of conditions (i) and (ii). This can be proved by the same method which was used in the second part of Laurent [4, Theorem 3.5.2]. □

We conclude this note with the following remarks.

Remark 2. It should be mentioned that the strictly increasing sequence $k_1,\ldots, k_{n+1}$ satisfying conditions (i) and (ii) of Proposition 3 is an ε-alternant in the sense of Bridges [1, p. 280].

Remark 3. If we suppose that ε = 0 then condition (ii) of Proposition 3 becomes $\mu(\bar{x}) = |e(k_i)|$, i.e., the error is maximum at the points $k_1,\ldots, k_{n+1}$. So the Chebyshev alternation theorem - see, for example, Laurent [4, Theorem 3.5.2] - can be considered as a special case of Proposition 3 above.

ACKNOWLEDGMENTS

The authors wish to acknowledge valuable discussions with Jean-Baptiste Hiriart-Urruty (Clermont-Ferrand II, now at Toulouse III) regarding this work.

REFERENCES

1. D.S. Bridges, On the foundations of best approximation theory , Journal of Approximation Theory 28 (1980), 273-292.

2. R.C. Buck, Applications of duality in approximation theory , in : H.L. Garabedian, ed., Approximation of functions (Elsevier, New York, 1965), pp. 27-42.

3. V.F. Dem'yanov and V.N. Malozemov, Introduction to minimax (Wiley, New York, 1974).

4. P.-J. Laurent, Approximation et Optimisation (Hermann, Paris, 1972).

5. R.T. Rockafellar, Convex Analysis (Princeton University Press, Princeton, NJ, 1970).

6. I. Singer, Best approximation in normed linear spaces by elements of linear subspaces (Springer-Verlag, Berlin, 1970).

7. J.-J. Strodiot, V.H. Nguyen and N. Heukemes, ε-Optimal solutions in nondifferentiable convex programming and some related questions , Mathematical Programming, in press.

8. M. Valadier, Sous-différentiels d'une borne supérieure et d'une somme continue de fonctions convexes, Comptes Rendus de l'Académie des Sciences de Paris 268 (1969), pp. 39-42.

II. ALGORITHMS

A MULTIPLE EXCHANGE ALGORITHM IN CONVEX PROGRAMMING

H.-P. Blatt*, U. Kaiser, B. Ruffer-Beedgen

Universität Mannheim
Mannheim, West Germany

ABSTRACT

An exchange algorithm is given which is a generalization of the Remez algorithm respectively the Cheney-Goldstein method. We follow an idea of Töpfer and consider at each step a finite sequence of finite subproblems. Modifying a strategy of Carasso and Laurent we replace in the well known exchange theorem the zero-checks by practical δ-checks with $\delta > 0$. This allows us to get convergence under weaker assumptions, to reduce the numerical calculations of the functional to be minimized, to raise the stability of the algorithm, and last not least to introduce multiple exchange methods known from the Remez algorithm.

I. THE MINIMIZATION PROBLEM

Let T be an arbitrary set and

$$a : T \to \mathbb{R}^n, \quad f : T \to \mathbb{R}$$

be two bounded mappings, V a subspace of $\mathbb{R}^n$ with $\dim V = d$, W an affine subspace parallel to V. Denoting by $\langle\cdot,\cdot\rangle$ the scalar product in $\mathbb{R}^n$, we consider the continuous convex functional

*Current affiliation: Katholische Universitat Eichstatt, Eichstatt, West Germany

$$g(x) := \sup_{t\in T} (\langle a(t),x\rangle - f(t)).$$

The problem consists in <u>minimizing g(x) with respect to $x\in W$</u>. We suppose that

$$\alpha := \inf_{x\in W} g(x)$$

is finite. Furthermore we may assume without loss of generality that the set

$$\tilde{A} := \{(a(t),\ f(t)) \mid t\in T\} \subset \mathbb{R}^{n+1}$$

is convex, and we define $A := \{a(t) \mid t\in T\}$.

Carasso and Laurent [3] proposed a dual method for this problem which uses some generalized notions of Stiefel [10].

II. REFERENCES AND STRING OF REFERENCES

DEFINITION. A subset $R = \{a(t_i) \mid 1\le i\le k+1\}\subset A$ is called a <u>V-reference</u> if
(α) there exist $\lambda_i > 0$ $(1\le i\le k+1)$ such that

$$\sum_{i=1}^{k+1} \lambda_i = 1 \text{ and } \sum_{i=1}^{k+1} \lambda_i a_i \in V^{\perp},$$

(β) R is minimal, i.e., there exists no proper subset of R such that (α) holds.
The numbers λ_i are uniquely determined and called the <u>characteristic numbers</u> of the V-reference R.

DEFINITION. $x\in W$ is called <u>solution of the V-reference R in W</u>, if

$$\min_{z\in W}\ \max_{1\le i\le k+1} (\langle a(t_i),z\rangle - f(t_i)) = \max_{1\le i\le k+1} (\langle a(t_i),x\rangle - f(t_i)).$$

For constructing such a solution one defines

$$h_R = \sum_{i=1}^{k+1} \lambda_i(\langle a(t_i),z\rangle - f(t_i))$$

for a fixed $z\in W$ and solves the linear system

$$\langle a(t_i),y\rangle = f(t_i) - \langle a(t_i),z\rangle + h_R$$

for $1\le i\le k+1$. Then $x = z+y$ is a solution of the V-reference R in W, in general not unique. h_R is called <u>deviation of the V-reference</u> R in W.

Consider now the following construction:
Let $R_1\subset A$ be a V-reference, W_1 the set of solutions of the V-reference R_1 in

W, h_1 the deviation of R_1 in W,

$$V_1 := [R_1]^{\perp} \cap V.$$

Then $\dim V_1 = \dim V - k_1$, if R_1 has $k_1 + 1$ elements, and W_1 is parallel to V_1.
We repeat the same construction for V_1, W_1 instead of V, W:
Let $R_2 \subset A$ be a V_1-reference, W_2 the set of solutions of the V_1-reference R_2 in W_1, h_2 the deviation of R_2 in W_1,

$$V_2 := [R_2]^{\perp} \cap V_1 .$$

Then $\dim V_2 = \dim V - k_1 - k_2$, if R_2 has $k_2 + 1$ elements, and W_2 is parallel to V_2, etc...
In this way we possibly can find a <u>string of references</u>

$$R = (R_1, R_2, \ldots, R_s)$$

with corresponding subspaces

$$V_1 \supset V_2 \supset \ldots \supset V_s$$

and parallel affine subspaces

$$W_1 \supset W_2 \supset \ldots \supset W_s,$$

and a <u>vector of deviations</u>

$$h = (h_1, h_2, \ldots, h_s)$$

such that

(1) R_i is a V_{i-1}-reference,

(2) $V_i = [R_i]^{\perp} \cap V_{i-1}$,

(3) h_i is the deviation of R_i in W_{i-1},

(4) $V_s = \{0\}$

$(V_o := V,\ W_o := W)$.

Such a string of references determines a unique point $x_R \in W$ ($\{x_R\} = W_s$), which is the <u>solution of the string R in W</u>.

DEFINITION. R is called <u>regular</u>, if each R_i has at least two elements.

If R is not regular we get a regular string by cancelling in R each R_j having only one element. Thereby the solution of the string is not changed.

III. THE δ-EXCHANGE PROCEDURE

Let $\delta \geq 0$ and

$$R = (R_1, \ldots, R_j, R_{j+1}, \ldots, R_s)$$

a string of references with vector

$$h = (h_1, \ldots, h_j, h_{j+1}, \ldots, h_s)$$

of deviations. There exist characteristic numbers $\lambda_i > 0$ for the V_{j-1}-reference

$$R_j = \{a(t_i) \mid 1 \leq i \leq k+1\}$$

such that $\sum_{i=1}^{k+1} \lambda_i \, a(t_i) \in V_{j-1}^{\perp}$ and $\dim [R_j]/V_{j-1}^{\perp} = k$.

Analogously there exist for the V_j-reference

$$R_{j+1} = \{a(\tilde{t}_i) \mid 1 \leq i \leq \ell + 1\}$$

characteristic numbers $\mu_i > 0$ such that

$$z_o := \sum_{i=1}^{\ell+1} \mu_i \, a(\tilde{t}_i) \in V_j^{\perp} \text{ and } \dim [R_{j+1}]/V_j^{\perp} = \ell. \quad (1)$$

Because of $V_j^{\perp} = [R_j] + V_{j-1}$ we can find real numbers w_i such that

$$z_o + \sum_{i=1}^{k} w_i \, a(t_i) \in V_{j-1}^{\perp}.$$

With $w_{k+1} = 0$ we get for $\theta \in \mathbb{R}$:

$$z_o + \sum_{i=1}^{k+1} (w_i - \theta\lambda_i) \, a(t_i) = z_o + \sum_{i=1}^{k+1} \lambda_i \left(\frac{w_i}{\lambda_i} - \theta\right) a(t_i) \in V_{j-1}^{\perp}.$$

We choose

$$\theta = \min_{1 \leq i \leq k+1} (w_i/\lambda_i)$$

and define

$$\kappa := 1 + \sum_{i=1}^{k+1} (w_i - \theta\lambda_i), \quad (2)$$

$$\rho_i := (w_i - \theta\lambda_i)/\kappa \quad \text{for } 1 \leq i \leq k+1,$$

$$\tilde{\mu}_i := \mu_i/\kappa \quad \text{for } 1 \leq i \leq \ell + 1.$$

Then at least one of the ρ_i is zero and

$$\operatorname{sgn} \rho_i = \operatorname{sgn} \lambda_i > 0 \text{ for } \rho_i \neq 0.$$

We define

$$I(\delta) := \{i \mid 1 \le i \le k+1,\ \rho_i \le \delta\},$$

$$C_o := \{a(t_i) \mid i \in I(\delta)\},$$

$$\hat{\mu}_1 := \tilde{\mu}_1 + \sum_{i \in I(\delta)} \rho_i,$$

$$\tilde{a} := (\tilde{\mu}_1\ a(\tilde{t}_1) + \sum_{i \in I(\delta)} \rho_i\ a(t_i))/\hat{\mu}_1,$$

$$\tilde{f} := (\tilde{\mu}_1\ f(\tilde{t}_1) + \sum_{i \in I(\delta)} \rho_i\ f(t_i))/\hat{\mu}_1.$$

Then there exists a point $\tilde{t} \in T$ such that

$$\tilde{a} = a(\tilde{t}) \text{ and } \tilde{f} = f(\tilde{t}).$$

Using the same arguments as Carasso, Laurent [4] we see that

$$\tilde{R}_j = R_j \cup R_{j+1} \cup \{a(\tilde{t})\} - C_o - \{a(\tilde{t}_1)\}$$

is a V_{j-1}-reference and $\tilde{R}_{j+1} := C_o$ a $\tilde{V}_j$-reference with $\tilde{V}_j = [\tilde{R}_j]^{\perp} \cap V_{j-1}$. Furthermore we get for the deviation $\tilde{h}_j$ of the V_{j-1}-reference $\tilde{R}_j$:

$$\tilde{h}_j = \kappa^{-1}\ h_{j+1} + (1-\kappa^{-1})h_j. \tag{3}$$

Summarizing we have got a new string

$$\tilde{R} = (R_1,\dots,R_{j-1},\tilde{R}_j,\tilde{R}_{j+1},R_{j+2},\dots,R_s)$$

with the properties:

1. $R_j = B_o \cup C_o$ with $C_o \neq \emptyset$,
2. $B_o \subset \tilde{R}_j$, $\tilde{R}_j \subset A$, $\tilde{R}_{j+1} = C_o$,
3. the characteristic numbers of $\tilde{R}_j$ belonging to elements of B_o are all $\ge \delta$,
4. $\tilde{h}_j \in (h_j, h_{j+1}]$ if $h_{j+1} > h_j$.

We remark that the deviation $\tilde{h}_j$ doesn't depend on δ. For $\delta = 0$ we get the exchange theorem of Carasso and Laurent ([3],[4]).

IV. THE EXCHANGE ALGORITHM

The algorithm depends on the parameters

$$\varepsilon > 0 \text{ and } \frac{1}{2} \le \beta < 1.$$

We want to construct an ε-solution of our problem, i.e., $z \in W$ such that

$$g(z) \le \alpha + \varepsilon.$$

We assume that we can find a string of references

$$R^1 = (R_1^1, \ldots, R_{s_1}^1)$$

with the solution x^1 and the vector of deviations

$$h_1 = (h_1^1, \ldots, h_{s_1}^1).$$

We put

$$K_1 := g(x^1)$$

and describe now <u>the step ν of the algorithm</u>:

We have a string of references

$$R^\nu = (R_1^\nu, \ldots, R_{s_\nu}^\nu)$$

with solution x^ν and vector of deviations

$$h^\nu = (h_1^\nu, \ldots, h_{s_\nu}^\nu),$$

a number $K_\nu \in \mathbb{R}$ and an index $1 \le \mu \le \nu$ such that

$$g(x^\mu) \le K_\nu.$$

We distinguish the cases:

(a) $h_1^\nu \ge K_\nu - \varepsilon$:

Then x^μ is an ε-solution and the algorithm stops.

(b) $h_1^\nu < K_\nu - \varepsilon$:

In this case we have to consider two cases again.

(b1) If there exists a <u>minimal</u> index j with $1 \le j \le s_\nu - 1$ such that

$$h_{j+1}^\nu \ge \beta h_j^\nu + (1-\beta)K_\nu,$$

we determine as in (2) the number $\kappa = \kappa_\nu$.

Let $k_\nu + 1$ be the number of points of R_j^ν. We define

$$\delta_\nu := \frac{\varepsilon}{(k_\nu+1)(2^\nu(K_\nu - h_j^\nu) - \varepsilon)}.$$

If $\kappa_\nu^{-1} \ge \delta_\nu$ we exchange R_j^ν and R_{j+1}^ν following the δ_ν-exchange of the preceding section. If $\kappa_\nu^{-1} < \delta_\nu$ we formally set up the new string of references

$$R = (R_1^\nu, \ldots, R_j^\nu, \{z_o\}, R_{j+1}^\nu, \ldots, R_{s_\nu}^\nu)$$

with z_o as in (1). Then we exchange R_j^ν and $\{z_o\}$ with the δ_ν-exchange method and put $K_{\nu+1} = K_\nu$.

(b2) If such a minimal j as in (b1) doesn't exist, we form R^ν to a regular string R, determine a point $t_o \in T$ such that

$$\langle a(t_o), x^\nu\rangle - f(t_o) \geq g(x^\nu) - \frac{\varepsilon}{2^{s_\nu}}$$

and put

$$K_{\nu+1} := \min (K_\nu, g(x^\nu)).$$

Now we define the new string of references by

$$R^{\nu+1} = (P, \{a(t_o)\}).$$

V. MULTIPLE EXCHANGE TECHNIQUES

Under certain conditions the Remez algorithm for the approximation of functions with respect to n-dimensional Haar subspaces has quadratic convergence if one exchanges not only one point with maximal deviation in the error curve, but n + 2 local extreme points of the error curve. We want to suggest such a proceeding to get a faster convergence.

We assume that the set T is a compact metric space. Then we can modify the iteration step of type (b2) in the following way:

($\tilde{b}$2) If a minimal j as in (b1) doesn't exist, we form R^ν to a regular string R and determine the relative maxima t of the function

$$\langle a(t), x^\nu\rangle - f(t)$$

in T. Let M be the set of these relative maxima and let $M = \{t_i \mid 1 \leq i \leq r\}$ be a finite set, then we define the new string by

$$R^{\nu+1} = (R, \{a(t_1)\}, \ldots, \{a(t\ )\})$$

and put

$$K_{\nu+1} = \min (K_\nu, g(x^\nu)).$$

VI. PROPERTIES OF THE ALGORITHM

First we shall show that the algorithm is well-defined and doesn't get into a cycle.

LEMMA 1. If the vector h^ν of deviations satisfies the inequality

$$h_j^\nu \geq K_\nu - \frac{\varepsilon}{2^{j-1}}$$

for an index $j \in \{1,\ldots,s_\nu\}$, then either $h_1^\nu \geq K_\nu - \varepsilon$ or there exists an index $i_o \in \{1,\ldots,j-1\}$ such that

$$h_{i_o+1}^\nu \geq \beta h_{i_o}^\nu + (1-\beta)K_\nu.$$

Proof. Let us assume that

$$h_{i+1}^\nu < \beta h_i^\nu + (1-\beta)K_\nu$$

for each $i \in \{1,\ldots,j-1\}$. Then

$$K_\nu - \frac{\varepsilon}{2^{j-1}} \leq h_j^\nu < \beta h_{j-1}^\nu + (1-\beta)K_\nu$$

and therefore we have

$$h_{j-1}^\nu > K_\nu - \frac{\varepsilon}{\beta \cdot 2^{j-1}}$$

$$\geq K_\nu - \frac{\varepsilon}{2^{j-2}}$$

because of $\beta \geq \frac{1}{2}$. Now the index $j-1$ satisfies the assumption of the Lemma and after a finite number of steps we get

$$h_1^\nu > K_\nu - \varepsilon. \quad \square$$

COROLLARY. In the algorithm a step of type (b2) (resp. ($\tilde{b}$2)) is always succeeded by a step of type (a) or (b1).

LEMMA 2. If the exchange index j at step ν satisfies

$$h_{j+1}^\nu \geq K_\nu - \frac{\varepsilon}{2^j},$$

then in the next step we have

$$h_j^{\nu+1} > K_\nu - \frac{\varepsilon}{2^{j-1}}$$

if $|R_{j}^{\nu+1}| = 1$.

Proof. Since $|R_{j}^{\nu+1}| = 1$ we know, following the exchange method, that $I(\delta_\nu) = \{i \mid 1 \le i \le k_\nu + 1\}$.

Therefore we have (see (3))

$$h_{j}^{\nu+1} = \kappa_\nu^{-1} h_{j+1}^{\nu} + (1-\kappa_\nu^{-1}) h_{j}^{\nu}$$

with

$$1 - \kappa_\nu^{-1} \le k_\nu \cdot \delta_\nu.$$

Then we can deduce using the definition of δ_ν:

$$\begin{aligned} h_{j}^{\nu+1} &\ge k_\nu \cdot \delta_\nu \cdot h_{j}^{\nu} + (1-k_\nu \cdot \delta_\nu) h_{j+1}^{\nu} \\ &\ge k_\nu \cdot \delta_\nu \cdot h_{j}^{\nu} + (1-k_\nu \cdot \delta_\nu)(K_\nu - \frac{\varepsilon}{2^j}) \\ &= K_\nu - \frac{\varepsilon}{2^j} - k_\nu \cdot \delta_\nu (K_\nu - \frac{\varepsilon}{2^j} - h_{j}^{\nu}) \\ &> K_\nu - \frac{\varepsilon}{2^{j-1}}. \quad \square \end{aligned}$$

COROLLARY. If the step ν of the algorithm is a step of type (b2) (resp. ($\tilde{b}$2)) then after a finite number of steps we get an ε-solution or a solution $x^{\nu_1} \ne x^{\nu}$ of a string R^{ν_1} with $\nu_1 > \nu$.

LEMMA 3. Let $k_\nu + 1$ be the number of points of R_{j}^{ν} and j the exchange index at step ν of type (b1) then the exchange parameter δ_ν satisfies

$$0 < \delta_\nu < \frac{1}{k_\nu + 1}.$$

Proof. Because of Lemma 1 and the minimal property of j we have

$$h_{j}^{\nu} < K_\nu - \frac{\varepsilon}{2^{j-1}}$$

and therefore

$$2^j (K_\nu - h_{j}^{\nu}) - \varepsilon > \varepsilon$$

which is equivalent to the assertion. □

COROLLARY. (a) The algorithm is well-defined. (b) If we have $\kappa_\nu^{-1} < \delta_\nu$ then the reference $R_j^{\nu+1}$ has at least two points.

Proof. Following the exchange procedure of section 3 we have for $\kappa_\nu^{-1} < \delta_\nu$:

$$1 = \sum_{i=1}^{k_\nu+1} \rho_i + \kappa_\nu^{-1}.$$

At least one of the numbers ρ_i is 0. Therefore we get from the above equality that at least one of the ρ_i is greater than δ_ν. Then $R_j^{\nu+1}$ has at least two points. □

VII. CONVERGENCE OF THE ALGORITHM

In general the vectors h^ν are not monotonically increasing in the lexicographical sense. But if we consider the regular string $\overline{R}^\nu$ belonging to R^ν with corresponding deviation vector $\overline{h}^\nu$ we can easily see that the sequence $\{\overline{h}^\nu\}$ increases monotonically in the lexicographical sense. Moreover we have for the convergence of the algorithm the following

THEOREM. After a finite number of steps the algorithm with single or multiple exchanges leads to an ε-solution.

Proof. Let us assume that the algorithm doesn't stop. Then we assert: There exists no natural number ν_o such that

$$\overline{h}^\nu = \overline{h}^{\nu_o} \text{ for all } \nu \geq \nu_o. \tag{4}$$

Let us assume the contrary. Then the corollary of Lemma 2 says, that we have no step of type (b2) (resp. ($\tilde{b}$2)) for $\nu \geq \nu_o$. Moreover in all steps ν with $\nu \geq \nu_o$ we always exchange R_{j+1}^ν with R_j^ν and R_{j+1}^ν contains only one point. Then we deduce from the Corollary of Lemma 3 that at each step ν ($\nu \geq \nu_o$) we have $\kappa_\nu^{-1} \geq \delta_\nu$. Therefore for all $\nu \geq \nu_o$ the strings $\overline{R}^\nu$ always have the same number of references. Now let $j(\nu)$ be the exchange index of step ν. At step ν_o we exchange $R_{j(\nu_o)}^{\nu_o}$ with $R_{j(\nu_o)+1}^{\nu_o}$ and after a finite number of steps we reach an index ν_1 such that

$$j(\nu_1) > j(\nu_o),$$

since the exchange index $j(\nu_o)$ is minimal. Using the same arguments, the algorithm leads to a finite index ν_k such that an exchange index $j(\nu_k)$ exists no more. Then we have at step ν_k a step of type (b2) (resp. ($\tilde{b}$2)), which is a contradiction. Therefore the assertion (4) is true. Since the vectors $\bar{h}^\nu$ are monotonically increasing, there exists a maximal natural number m and an index ν_o such that

$$\bar{h}_j^\nu = \bar{h}_j^{\nu+1} \quad \text{for } 1 \le j \le m-1 \text{ and } \nu \ge \nu_o.$$

Because of (4) there exists a subsequence $\{\nu_k\}$ of $\mathbb{N}$ such that

$$\nu_o \le \nu_1 < \nu_2 < \ldots \tag{5}$$

and

$$\bar{R}_m^{\nu_k} \neq \bar{R}_m^{\nu_k+1} = \bar{R}_m^{\nu_k+2} = \cdots = \bar{R}_m^{\nu_{k+1}} \tag{6}$$

for $k = 1,2,\ldots$. Now we use

LEMMA 4. $\overline{\lim_{k\to\infty}}\ \kappa_{\nu_k}^{-1} = 2\tau > 0.$

Therefore there exists a subsequence $\{\mu_k\}$ of $\{\nu_k\}$ with $\lim_{k\to\infty} \kappa_{\mu_k}^{-1} = 2\tau$. From (3) we get

$$\bar{h}_m^{\mu_k+1} \ge \tau\, \bar{h}_{m+1}^{\mu_k} + (1-\tau)\bar{h}_m^{\mu_k}$$

with

$$\bar{h}_{m+1}^{\mu_k} \ge \beta\, \bar{h}_m^{\mu_k} + (1-\beta)K_{\mu_k}$$

for all sufficiently large $k \ge k_o$. Now we choose $\tilde{\mu}_k \le \mu_k$ such that

$$g(x^{\tilde{\mu}_k}) \le K_{\mu_k}.$$

Then we get

$$\bar{h}_m^{\mu_k+1} \ge (1-\tau(1-\beta))\bar{h}_m^{\mu_k} + \tau(1-\beta)g(x^{\tilde{\mu}_k})$$

and

$$\bar{h}_m^{\mu_k+1} - \bar{h}_m^{\mu_k} \ge \tau(1-\beta)[g(x^{\tilde{\mu}_k}) - \bar{h}_m^{\mu_k}].$$

Since $\{\bar{h}_m^\nu\}$ is monotonically increasing and bounded, we can deduce that

$$\lim_{\nu\to\infty} \bar{h}_m^\nu = \lim_{k\to\infty} g(x^{\tilde{\mu}_k}).$$

Since the exchange index is always minimal, it follows that

$$\lim_{\nu\to\infty} \bar{h}_1^\nu = \lim_{k\to\infty} g(x^{\tilde{\mu}_k}) = \alpha.$$

Therefore there exists an index $\tilde{\nu} \in \mathbb{N}$ such that

$$\bar{h}_1^{\tilde{\nu}} > K_{\tilde{\nu}} - \varepsilon$$

and the algorithm stops at step $\tilde{\nu}$. This contradicts our assumption and the theorem is proved. □

Now we have to add the proof of Lemma 4.

Proof of Lemma 4. Let us assume that

$$\lim_{k\to\infty} \kappa_{\nu_k}^{-1} = 0. \tag{7}$$

We know that

$$\delta_{\nu_k} = \frac{\varepsilon}{(\ell_k+1)(2^m(K_{\nu_k} - h_m^{\nu_k}) - \varepsilon)}$$

$$\geq \frac{\varepsilon}{(n+1)(2^m(K_{\nu_1} - h_m^{\nu_1}) - \varepsilon)}$$

$$=: \delta$$

($\ell_k + 1 = |R_m^{\nu_k}|$). Then $\delta > 0$, and we assume that for all $k = 1,2,\ldots$ the inequality

$$\kappa_{\nu_k}^{-1} < \delta$$

holds. Therefore at each step $\nu_k (k = 1,2,\ldots)$ we have an exchange between $\bar{R}_m^{\nu_k}$ and another reference having only one element. Then in all references $\bar{R}_m^{\nu_k}$ $(k = 1,2,\ldots)$ we have the same number $r + 1$ of elements $(r \geq 1)$. Now let us assume that in

$$\bar{R}_m^\nu = \{a_i^\nu \mid 0 \leq i \leq r\}$$

the element a_r^ν is introduced at step $\nu = \nu_k$ into the reference $\bar{R}_m^{\nu_k}$. We define

$$b_i^\nu := p(a_i^\nu),$$

denoting by p the orthogonal projection of $\mathbb{R}^n$ into $\overline{V}_{m-1}^\nu$. We determine the characteristic numbers of the $\overline{V}_{m-1}^\nu$-reference $\overline{R}_m^\nu$ by the following way: Let

$$\lambda_i^\nu = \sqrt{G(b_o^\nu,\ldots,b_{i-1}^\nu,b_{i+1}^\nu,\ldots,b_r^\nu)},$$

$$\lambda = \sum_{i=o}^{r} \lambda_i^\nu,$$

$$\tilde{\lambda}_i^\nu = \lambda_i^\nu / \lambda,$$

denoting by $G(x_1,\ldots,x_r)$ the Gram determinant of the vectors $x_1,\ldots,x_r$. Then the numbers $\tilde{\lambda}_i^\nu$ $(0 \le i \le r)$ are the characteristic numbers of the $\overline{V}_{m-1}^\nu$-reference $\overline{R}_m^\nu$. Therefore

$$\kappa_{\nu_k}^{-1} = \tilde{\lambda}_r^{\nu_k+1} \tag{8}$$

and, following the exchange procedure of the algorithm, we know that

$$0 < \tilde{\lambda}_r^\nu < \tilde{\lambda}_i^\nu$$

for all $0 \le i \le r-1$. Since

$$\lambda_r^{\nu+1} \in \{\lambda_o^\nu,\ldots,\lambda_r^\nu\}$$

the sequence $\{\lambda_r^\nu\}$ is monotonically increasing. On the other hand from (7), (8) and the boundedness of A it follows that

$$\lim_{\nu\to\infty} \lambda_r^\nu = 0.$$

This is a contradiction and the Lemma is proved. □

VIII. APPLICATION TO APPROXIMATION PROBLEMS

Let E be a normed linear space over $\mathbb{C}$ with norm $\|\cdot\|$, $V = \text{span}(v_1,v_2,\ldots,v_n)$ a n-dimensional subspace of E. For a given element $f \in E$ we want to find an element $\overline{v} \in V$ such that

$$\min_{v\in V} \| f-v \| = \| f-\overline{v} \|. \tag{9}$$

If S_{E*} denotes the unit cell of the dual space E* it is known that

$$\| h \| = \max_{L\in S_{E*}} \text{Re } L(h).$$

Consider for $v \in V$ the representation

$$v = \sum_{k=1}^{n} (x_k + i x_{k+n}) v_k$$

and define

$$x := (x_1, \ldots, x_{2n}),$$

$$a(L) := (\mathrm{Re}L(v_1), \ldots, \mathrm{Re}L(v_n), -\mathrm{Im}L(v_1), \ldots, -\mathrm{Im}L(v_n)),$$

$$f(L) := \mathrm{Re}L(f).$$

Then

$$\mathrm{Re}L(v-f) = \langle a(L), x\rangle - f(L)$$

and the problem (9) is equivalent to minimizing the functional

$$g(x) = \max_{L \in S_{E^*}} \langle a(L), x\rangle - f(L)$$

with respect to $x \in \mathbb{R}^{2n}$.

Remark. In this case the set $A = \{a(L) \mid L \in S_{E^*}\}$ satisfies $A = -A$. Therefore it is possible to modify slightly the algorithm to get at each step a deviation vector which is nonnegative.

IX. EXAMPLES

(a) We want to determine an approximation for the conformal mapping f of the unit square $B = [-1,1] \times [-1,1]$ to a circle $S = \{z \mid |z| < \rho\}$ under the constraints $f(0) = 0$ and $f'(0) = 1$. It is known (Golusin [6]) that f is characterized by

$$\sup_{z \in B} |f(z)| = \min_{h \in H(B)} \sup_{z \in B} |h(z)|,$$

where H(B) denotes the set of all functions h holomorphic in B with $h(0) = 0$ and $h'(0) = 1$. To get an approximation of f we consider the problem (P):

Minimize the functional $\max_{z \in \partial B} |p(z)|$ with respect to the set of all polynomials p(z) of degree not greater than n satisfying $p(0) = 0$, $p'(0) = 1$. We restrict n to 17. Because of the symmetry of the problem we know that the solution of (P) has the form

$$p(z) = z + \alpha_1 z^5 + \alpha_2 z^9 + \alpha_3 z^{13} + \alpha_4 z^{17}$$

with $\alpha_i \in \mathbb{R}$ for $1 \le i \le 4$. Moreover we can restrict the approximation to the set

$$B_o = \{z = x + i \mid x \in [0,1]\}.$$

To simplify the calculation of the error this set B_o was replaced by a discrete set $B_1 = \{x_1 < x_2 < \ldots < x_{200}\}$ of 200 points uniformly distributed on B_o. The five starting points were

$$z_1 = x_1,\ z_2 = x_{50},\ z_3 = x_{100},\ z_4 = x_{150},\ z_5 = x_{200}$$

and we begin the first step of the algorithm with the $\mathbb{R}^4$-reference

$$R = \bigcup_{j=1}^{5} a(L_j) \text{ with } L_j(v_i) = |v_i(z_j)| \text{ for } 1 \le i \le 4.$$

We put $\beta = 0.5$ and $\varepsilon = 10^{-8}$ and get the results given in table 1 for the algorithm with single exchange, respectively multiple exchange.

TABLE 1: Problem (a)

step	single exchange $g(x^m) - h_1^m$	multiple exchange $g(x^m) - h_1^m$
1	0.638312 (-2)	In the first three steps
2	0.634898 (-2)	only the single exchange
3	0.517130 (-2)	method was used
4	0.325266 (-2)	0.31888 (-3)
5	0.32848 (-3)	0.2034 (-4)
6	0.2081 (-4)	0.481 (-5)
7	0.2209 (-4)	0.166 (-5)
8	0.3580 (-4)	0.44 (-6)
9	0.547 (-5)	0.14 (-6)
10	0.3242 (-4)	0.3 (-7)
11	0.106 (-5)	$< 10^{-8}$
12	0.30 (-6)	
13	0.25 (-6)	
14	0.7 (-7)	
15	0.6 (-7)	
16	0.1 (-7)	
17	$< 10^{-8}$	

It should be mentioned that in table 1 the number of steps is equivalent to the number of calculations of the functional g(x), which is the most time consuming part of the algorithm.

If $g(x^m) - h_1^m < 10^{-8}$ we get an approximation $p_1(z)$ of the conformal mapping f:

$$p_1(z) = z + 0.0738653\ z^5 + 0.0045426\ z^9 + \\ + 0.0002806\ z^{13} + 0.0000137\ z^{17},$$

$$\max_{z \in B_1} |p_1(z)| = 1.07872293.$$

(b) We approximate the function $f(x,y) = x^2y$ on the unit disc with respect to the subspace

$$V = \mathrm{span}(1, x, x^2, y, y^2, xy)$$

in the Chebyshev-norm. It is known that the set of best approximations is represented by

$$r(x,y) = 0.25y + \gamma(x^2+y^2-1)$$

with $-0.125 \leq \gamma \leq 0.125$ (Andreassen, Watson [1]). Starting with an equidistant point set on the unit circle, $\varepsilon = 10^{-8}$ and $\beta = 0.9$ the single exchange method reaches an exact solution at step 9. By comparison using the multiple exchange after the third step we get already an exact solution at step 5. The final result for the parameter γ and the deviation vector $h = (h_1, h_2)$ are listed in table 2.

TABLE 2: Problem (b)

	single exchange	multiple exchange
γ	0.059	0.034
h_1	0.25	0.25
h_2	0.1594...	0.1547...

(c) The function $f(x) = \sqrt{x}$ was approximated on [0,2] by cubic splines with knots at

$$x_1 = 0.001,\ x_2 = 0.015,\ x_3 = 0.06,\ x_4 = 0.2,\ x_5 = 0.35$$

(see also Nürnberger, Sommer [8]). With $\beta = 0.5$ and $\varepsilon = 10^{-10}$ the single exchange method stopped at step 16, whereas the multiple exchange needed 11 steps.

A lot of further numerical problems in real and complex Chebyshev-approximation and L_1-approximation was treated by these algorithms in [7], [9], [12].

Naturally the speed of convergence depends on the problem to be treated. In any case, if it is possible to introduce multiple exchange techniques, we have observed that the multiple exchange is superior to the single exchange. Furthermore it was possible to get suitable approximations to be improved by faster local methods such as Newton's method, even if the solution of the problem was extremely bad conditioned. Here bad condition means that the solution, as in the L_1-case, is characterized by a trivial reference consisting only of one element.

Finally we want to mention that we are able to get error estimates for the parameters of the solutions using the strings and the characteristic numbers of the associated references [2].

REFERENCES

1. D.O. Andreassen, G.A. Watson, Linear Chebyshev approximation without Chebyshev sets, BIT 16 (1976), 349-362.

2. H.-P. Blatt, Strenge Eindeutigkeitskonstanten und Fehlerabschätzungen bei linearer Tschebyscheff-Approximation, to appear in L. Collatz, G. Meinardus, H. Werner (Eds.): Numerische Methoden der Approximationstheorie, Birkhäuser Verlag, Basel, Tagung Oberwolfach 1981.

3. C. Carasso, P.J. Laurent, Un algorithme de minimisation en chaine en optimisation convexe, Siam J. Control and Optimization 16 (1978), 209-235.

4. C. Carasso, P.J. Laurent, An algorithm of successive minimization in convex programming, R.A.I.R.O., Analyse numérique, Numerical Analysis (1978), 377-400.

5. E.W. Cheney, A.A. Goldstein, Newton's method for convex programming and Tchebycheff approximation, Numerische Mathematik 1 (1959), 253-268.

6. G.M. Golusin, Geometric Theory of Functions of a Complex Variable, Translation of Mathematical Monographs, Vol. 26, Amer. Math. Soc., Providence, Rhode Island, 1969.

7. U. Kaiser, Ein Aufstiegsverfahren für konvexe Optimierungsprobleme mit Nebenbedingungen, Diplomarbeit, Universität Mannheim, 1981.

8. G. Nürnberger, M. Sommer, A Remez type algorithm for spline functions, Bericht Nr. 062 des Instituts für Angewandte Mathematik der Universität Erlangen-Nürnberg, April 1981.

9. B. Ruffer-Beedgen, Der Rémèz-Algorithmus ohne Haarsche Bedingung, Diplomarbeit, Universität Mannheim, 1981

10. E. Stiefel, Über diskrete und lineare Tschebyscheff-Approximation, Num. Math. 1 (1959), 1-28.

11. H.J. Töpfer, Tschebyscheff-Approximation und Austauschverfahren bei nicht erfüllter Haarscher Bedingung, ISNM 7, Birkhäuser Verlag, Basel 1967, 71-89.

12. J. Witte, Remes- und Newton-Verfahren bei L_1-Approximation, Staatsexamensarbeit, Universität Mannheim, 1981.

ALGORITHMES POUR EXTRAIRE UNE SOUS-SUITE CONVERGENTE D'UNE SUITE NON CONVERGENTE

J.-P. Delahaye

University of Lille I
U.E.R d'I.E.E.A.
Villeneuve d'Ascq, France

INTRODUCTION

Dans bien des questions liées à l'optimisation, on rencontre des suites non convergentes (algorithmes généraux ; optimisation non différentiable ; algorithmes de recherche de points fixes). Il a donc semblé intéressant de définir et d'étudier des algorithmes qui, à partir d'une suite non convergente (donnée progressivement) extraient des sous-suites convergentes.

Deux points de vue peuvent être adoptés :

1) On essaye d'extraire simultanément une famille de sous-suites convergentes telle qu'à chaque point d'accumulation de (x_n) corresponde au moins une sous-suite extraite.

2) On essaye d'extraire une seule sous-suite convergente.

Le premier point de vue est considéré dans [2][4]. Nous nous intéressons ici uniquement au second.

Notre travail consiste en la description et l'étude de deux types d'algorithmes permettant d'extraire une sous-suite. Ces deux types d'algorithmes sont basés sur un principe très simple qui est le suivant :

a) on découpe la suite (x_n) en tranches

$\{x_o, x_1, \ldots, x_{\beta(o)}\}$ tranche 0

$\{x_{\alpha(1)}, x_{\alpha(1)+1}, \ldots, x_{\beta(1)}\}$ tranche 1; $\alpha(1) \leq \beta(0)+1$

$\{x_{\alpha(2)}, x_{\alpha(2)+1}, \ldots, x_{\beta(2)}\}$ tranche 2; $\alpha(2) \leq \beta(1)+1$

etc ...

b) on prend un point dans chacune des tranches de ce découpage et on obtient ainsi une sous-suite extraite notée (t_i) ou (s_i).
Dans le premier type d'algorithmes (algorithmes T), on choisit t_{i+1} de façon à rendre $d(t_{i+1}, t_i)$ petit. Dans le second type d'algorithmes (algorithmes S), on choisit s_i de façon à rendre $d(s_i, a)$ petit (a est un point fixé).
Nos résultats établissent des conditions suffisantes de convergence pour les algorithmes T ou S. Ils montrent, entre autres qu'il existe toujours des découpages (c'est à dire des suites $\alpha(n)$, $\beta(n)$) pour lesquels T et S sont convergents). Ils indiquent aussi pour certains types particuliers de suites comment choisir $(\alpha(n), \beta(n))$.

Nous concluons en établissant qu'aucun algorithme ne peut extraire une sous-suite convergente de toute suite non convergente, et donc qu'en particulier aucun découpage $(\alpha(n), \beta(n))$ ne pourra fonctionner quelle que soit la suite (x_n).

I. NOTATIONS ET RAPPELS

Nous utiliserons les notations suivantes :

$\mathbb{N}$ (resp. $\mathbb{N}^*$) : ensemble des entiers ≥ 0 (resp > 0) ;
$\mathbb{R}$ (resp. $\mathbb{R}^+$, $\mathbb{R}^{+*}$) : ensemble des réels (resp. ≥ 0, > 0) ;
$\mathbb{R}^n$: espace euclidien de dimension n ;
E : partie fermée de $\mathbb{R}^n$;
d(x,y) : distance entre deux points de E (nous ne supposons pas que d est la distance euclidienne) ;
d(A, B) : $\inf \{d(x, y) \mid x \in A, y \in B\}$;
B(x, r) : $\{y \mid d(x, y) < r\}$
$\delta(A, B)$: $\max \{\sup\{d(A, y) \mid y \in B\}, \sup \{d(x, B) \mid x \in A\}\}$
(distance de Hausdorff entre deux ensembles voir [1]) ;

(x_n) : suite infinie de points de E ;

$S(E)$: ensemble des suites infinies de points de E ;

$A(x_n)$: ensemble des points d'accumulation de la suite (x_n),

$\hat{x} \in A(x_n) \iff (\forall\varepsilon > 0,\ \forall n_o \in \mathbb{N},\ \exists n \geq n_o : d(x_n, \hat{x}) \leq \varepsilon)$;

$A(x_n)$ est toujours un fermé, et puisqu'ici E est à base dénombrable de voisinages, on sait que $\hat{x} \in A(x_n)$ si et seulement si il existe une sous-suite (x_{i_n}) convergeant vers $\hat{x}$.

Lorsque la suite (x_n) est fixée, on notera :

$$X_n^p = \{x_n, x_{n+1}, \ldots, x_p\} \quad n \leq p.$$

Si (A_n) est une suite d'ensembles on notera :

$$\operatorname*{Lim\,inf}_{n\to\infty} A_n = \{\hat{x} \mid \forall\varepsilon > 0,\ \exists n_o,\ \forall n \geq n_o,\ \exists x \in A_n : d(x, \hat{x}) \leq \varepsilon\}$$

$$\operatorname*{Lim\,sup}_{n\to\infty} A_n = \{\hat{x} \mid \forall\varepsilon > 0,\ \forall n_o,\ \exists n \geq n_o,\ \exists x \in A_n : d(x,\hat{x}) \leq \varepsilon\}$$

Lorsque $\operatorname*{Lim\,inf}_{n\to\infty} A_n = \operatorname*{Lim\,sup}_{n\to\infty} A_n = A$ on dit que (A_n) converge vers A et on note $\lim_{n\to\infty} A_n = A$ (pour plus de détails sur ces notions voir [1] [5]).

II. DESCRIPTION DES ALGORITHMES T ET EXEMPLES

En plus de la suite (x_n) de points de E, on suppose données dans tout le reste du texte des suites d'entiers $\alpha(n)$, $\beta(n)$ vérifiant les conditions suivantes.

$$(\alpha\beta) \quad \begin{cases} \alpha(0) = 0 \\ \forall n \in \mathbb{N} \quad \alpha(n+1) > \alpha(n) \\ \qquad\qquad \beta(n) \geq \alpha(n+1)-1 \end{cases}$$

Ces conditions impliquent bien sûr que :

$$\lim_{n\to\infty} \alpha(n) = \lim_{n\to\infty} \beta(n) = +\infty \text{ et } \forall n \in \mathbb{N} \quad \beta(n) \geq \alpha(n)$$

Les suites $(\alpha(n), \beta(n))$ nous permettent de découper la suite (x_n) en "tranches".

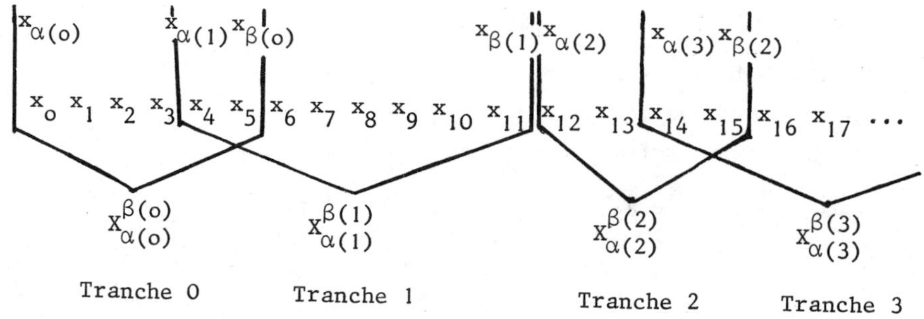

Figure 1.

Les sous-suites construites par l'algorithme T seront obtenues en prenant 1 point dans chacune des tranches.
Soit $i_o \in \mathbb{N}$.

ALGORITHME $T(i_o, \alpha(p), \beta(p))$.

Etape 0. On pose $t_o = x_{i_o}$. Soit p_o le plus petit entier tel que :

$$\alpha(p_o) \leq i_o \leq \beta(p_o)$$

Etape j. Soit I_j l'ensemble des entiers ℓ tels que :

$$\begin{cases} \alpha(p_o + j) \leq \ell \leq \beta(p_o + j) \\ d(x_\ell, t_{j-1}) = \min \; d(x_m, t_{j-1}) \mid \alpha(p_o + j) \leq m \leq \beta(p_o + j)\} \end{cases}$$

Soit $i_j = \max I_j$. On pose $t_j = x_{i_j}$.

Une fois les paramètres i_o, $(\alpha(n), \beta(n))$ fixés, l'algorithme T extrait une sous-suite (t_j) commençant par $t_o = x_i$, en prenant un point dans chacune des tranches définies par $(\alpha(n), \beta(n))$ et en minimisant la distance de deux points successifs de la sous-suite construite.

Nous prenons $i_j = \max I_j$ pour les cas où I_j a plus d'un élément mais ce choix n'est pas essentiel et les résultats établis resteraient valables avec une autre convention de choix (pourvu que $i_j \in I_j$).

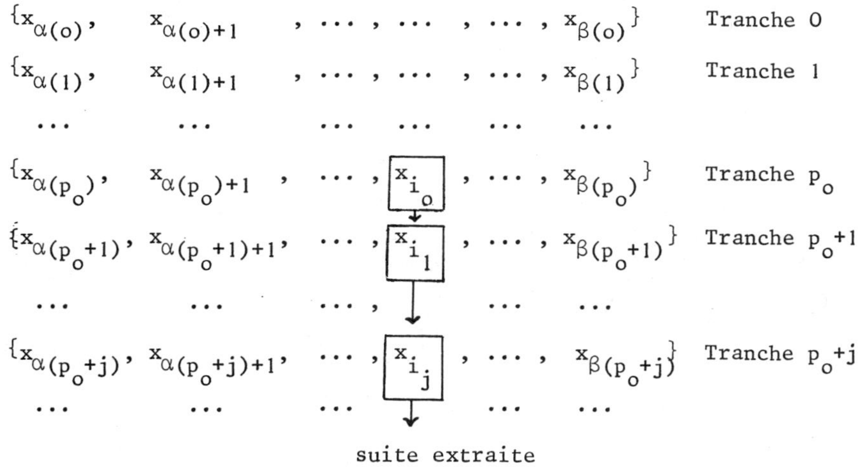

Figure 2.

Si la sous-suite $(x_{i_j}) = (t_j)$ converge vers $\hat{x}$, on écrira :

$$(i_o,\ \alpha(n),\ \beta(n)) \downarrow \hat{x}$$

ou même : $i_o \downarrow \hat{x}$

Pour exprimer que $(x_{i_j}) = (t_j)$ converge, on écrira :

$$(i_o,\ \alpha(n),\ \beta(n))\downarrow$$

ou même :

$$i_o\downarrow$$

<u>Exemple 1</u>. Prenons $E = \mathbb{R}$, et considérons (x_n) définie par :

$$\begin{cases} x_{3n} = 1/2^n \\ x_{3n+1} = 1+1/2^n \\ x_{3n+2} = 2+1/2^n \end{cases}$$

Fixons dans un premier temps $\alpha(n) = 3n$, $\beta(n) = 3n+2$.

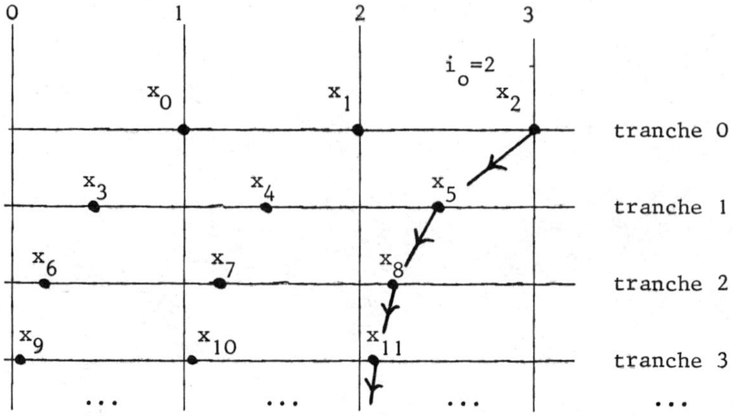

Figure 3.

Si on prend $i_o = 0$, on obtient la suite extraite $(x_o, x_4, x_7, x_{10}, \ldots)$ qui converge vers $1 \in A(x_n)$.

Ce que nous notons : $$i_o = 0 \downarrow 1.$$

De même : $$i_o = 1 \downarrow 2$$
$$i_o = 2 \downarrow 2$$
$$i_o = 3 \downarrow 0$$

Il est évident pour cet exemple que :

(*) $$\forall i_o \in \mathbb{N} \quad i_o{}^{\downarrow}$$

et que :

(**) $$\forall \hat{x} \in A(x_n),\ \exists i_o \in \mathbb{N} : i_o \downarrow \hat{x}$$

Fixons maintenant $\alpha(n) = 4n$, $\beta(n) = 4n+3$

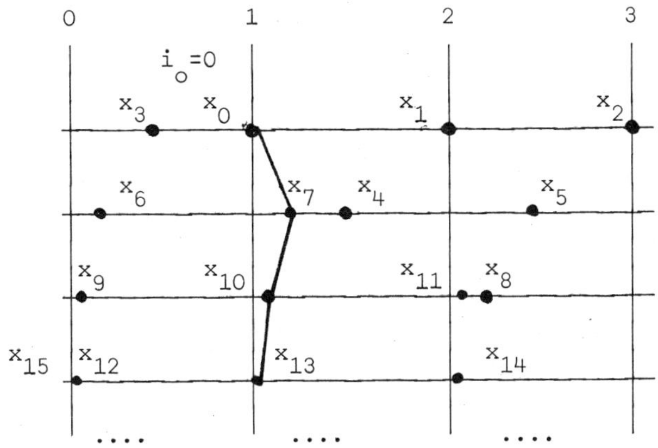

Figure 4.

Avec ce nouveau découpage, on a :

$$i_o = 0 \downarrow 1$$
$$i_o = 1 \downarrow 2$$
$$i_o = 2 \downarrow 2$$
$$i_o = 3 \downarrow 0$$

Les propriétés (*) et (**) sont encore satisfaites.
Sur cet exemple, on peut d'ailleurs établir sans difficulté qu'elles seront satisfaites dès que :

$$\forall n \in \mathbb{N} : \alpha(n+1) \geq \alpha(n) + 3$$

Exemple 2. Toujours avec $E = \mathbb{R}$, considérons maintenant :

$$(x_n) = (0, 0, 1/2, 0, 1/4, 1/2, 3/4, 0, 1/8, 1/4, 3/8, 1/2, 5/8, 3/4, 7/8, \ldots)$$

Il est clair que :

$$A(x_n) = [0, 1]$$

Pour que l'algorithme T extraie effectivement des sous-suites convergentes, il faut prendre des tranches de plus en plus longues. Par exemple, on peut fixer : $\alpha(p) = 2^p-1$, $\beta(p) = 2^{p+1}-2$

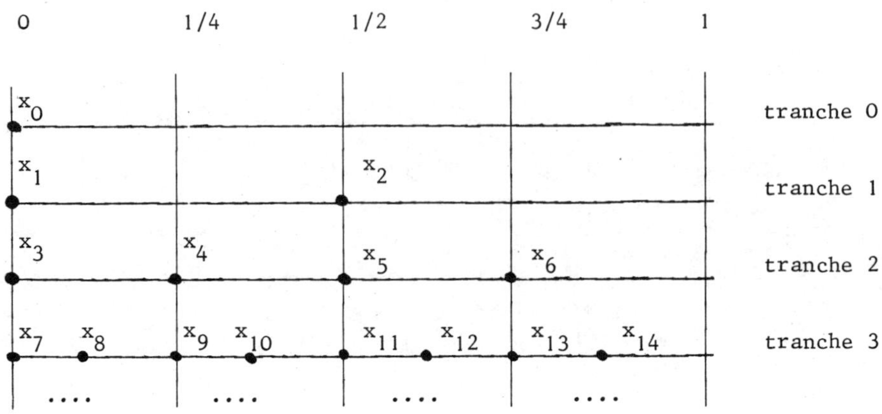

Figure 5.

On voit que (*) est encore vraie, par contre (**) n'est plus vraie.
Les points $\hat{x} \in A(x_n)$ tels que $\exists i_o : i_o \downarrow \hat{x}$, peuvent être caractérisés ; ce sont les points de la forme $\frac{k}{2^i}$ avec k impair, $0 \leq k \leq 2^i-1$.
Les points limites des sous-suites extraites par T ne peuvent constituer qu'un sous-ensemble dénombrable de $A(x_n)$ car les i_o possibles sont dénombrables) et ceci même en prenant des tranches plus grandes.
Remarquons qu'avec la suite de l'exemple 2, si $\beta(n) - \alpha(n)$ est borné, alors quel que soit le choix de i_o, la suite $(x_{i_j}) = (t_j)$ ne convergera pas (et même, sera telle que $A(t_j) = [0, 1]$). Prendre des tranches grandes est donc, pour de telles suites, une nécessité absolue.

III. THEOREMES DE CONVERGENCE POUR T

Les différents exemples que nous venons de voir montrent que l'algorithme T ne converge pas toujours, et que, lorsqu'il converge, il ne donne pas forcément n'importe quel point $\hat{x} \in A(x_n)$.
Pour obtenir des sous-suites convergentes, il est donc nécessaire de bien choisir $(\alpha(n), \beta(n))$.

Théorème 1. Si la condition suivante est vérifiée :

$$(\sigma) \quad \sum_{i \in \mathbb{N}} \delta(X_{\alpha(i)}^{\beta(i)}, X_{\alpha(i+1)}^{\beta(i+1)}) < \infty$$

alors :

(i) pour tout $i_o \in \mathbb{N}$: $i_o \downarrow$ (c'est à dire : l'algorithme T fournit toujours une sous-suite convergente).

(ii) pour tout point $\hat{x}$ isolé dans $A(x_n)$, il existe $i_{\hat{o}} \in \mathbb{N}$ tel que : $i_o \downarrow \hat{x}$.

Remarques. 1) La condition (σ) ne fait intervenir que la suite (x_n) (et pas $A(x_n)$ comme certaines des conditions rencontrées plus loin).

2) Les hypothèses de convergence pour les algorithmes d'extraction simultanée [2] [3] [4] excluaient les cas où $A(x_n)$ est infini. Ici notre hypothèse n'est pas restrictive pour $A(x_n)$, et il est tout à fait possible (comme dans l'exemple 2) de considérer une suite (x_n) ayant une infinité de points d'accumulation.

Démonstration du théorème 1. (i) Par définition de T :

$$d(t_i, t_{i+1}) = \min \{d(t_i, x) \mid x \in X_{\alpha(i+1)}^{\beta(i+1)}\}$$

$$= d(t_i, X_{\alpha(i+1)}^{\beta(i+1)})$$

Donc :

$$d(t_i, t_{i+1}) \le \sup \{d(y, X_{\alpha(i+1)}^{\beta(i+1)}) \mid y \in X_{\alpha(i)}^{\beta(i)}\}$$

$$\le \delta(X_{\alpha(i)}^{\beta(i)}, X_{\alpha(i+1)}^{\beta(i+1)})$$

Ce qui donne : $\sum_{i \in \mathbb{N}} d(t_i, t_{i+1}) < \infty$; la suite (t_i) est donc une suite de Cauchy et par conséquent converge.

(ii) Soit $\varepsilon = d(\hat{x}, A(x_n) - \{\hat{x}\}) > 0$. Soit i_o tel que :

$$\sum_{i \ge i_o} \delta(X_{\alpha(i)}^{\beta(i)}, X_{\alpha(i+1)}^{\beta(i+1)}) < \varepsilon/3 \text{ et } x_{i_o} \in B(\hat{x}, \varepsilon/3).$$

La suite (t_i) converge (d'après (i)) vers une limite $\hat{y} \in A(x_n)$. On a nécessairement :

$$d(x_{i_o}, \hat{y}) < 2\varepsilon/3$$

donc $\hat{y} = \hat{x}$.

Le théorème 1 suggère immédiatement la question : est-il toujours possible de trouver $(\alpha(n), \beta(n))$ satisfaisant (σ) (et ($\alpha\beta$)) ? La réponse à cette question est donnée ici :

Proposition 1. Soit (x_n) une suite bornée de points de E. Il existe des suites d'entiers $(\alpha(n), \beta(n))$ telles que (σ) soit vérifiée (ainsi que $(\alpha\beta)$).

Démonstration. Pour tout $\varepsilon > 0$, nous noterons :

$$V_\varepsilon = \{x \mid d(x, A(x_n)) < \varepsilon\}$$

Puisque (x_n) est une suite bornée de E, $A(x_n)$ est compact et on a donc

$$\forall\varepsilon > 0,\ \exists n(\varepsilon),\ \forall m \geq n(\varepsilon) : x_m \in V_\varepsilon.$$

Donnons-nous $(\varepsilon_i)_{i\geq 1}$, une suite de paramètres deux à deux distincts tels que :

$$\varepsilon_i > 0,\ \sum_{i\geq 1} \varepsilon_i < \infty.$$

On définit : $\alpha(0) = 0$, $\beta(0) = n(\varepsilon_1)$, $\alpha(1) = \beta(0)+1$.
$A(x_n)$ qui est compact, peut être recouvert à l'aide d'un nombre fini de boules de rayon $\varepsilon_1/2$:

$$\begin{cases} B(y_o^1, \varepsilon_1/2), \ \ldots\ , B(y_{\ell_1}^1, \varepsilon_1/2) \\ y_o^1, \ldots, y_{\ell_1}^1 \in A(x_n) \end{cases}$$

Prenons k_1 tel que pour tout $i = 0, \ldots, \ell_1$ il existe $n \in \{\alpha(1),\ldots,k_1\}$ tel que $x_n \in B(y_i^1, \varepsilon_1/2)$. On définit : $\beta(1) = \max\{k_1, \alpha(1), n(\varepsilon_2)\}$, $\alpha(2) = \beta(1)+1$. Montrons que :

$$(*) \qquad \delta(X_{\alpha(1)}^{\beta(1)}, A(x_n)) \leq \varepsilon_1.$$

Soit $x \in X_{\alpha(1)}^{\beta(1)}$, on a $d(x, A(x_n)) \leq \varepsilon_1$ (car $\alpha(1) \geq n(\varepsilon_1)$).
Soit $\hat{x} \in A(x_n)$, il existe y_r^1 tel que : $d(\hat{x}, y_r^1) \leq \varepsilon_1/2$ (par hypothèse sur les y_1^1). Il existe donc :

$$n \in \{\alpha(1),\ldots,\beta(1)\} \text{ tel que : } d(y_r^1, x_n) \leq \varepsilon_1/2 \ ;$$

donc tel que : $\quad d(\hat{x}, x_n) \leq \varepsilon_1.$
Ce qui signifie que :

$$d(\hat{x}, X_{\alpha(1)}^{\beta(1)}) \leq \varepsilon_1.$$

Ce qui montre (*).
On recouvre alors $A(x_n)$ à l'aide de boules de rayon $\varepsilon_2/2$:

$$\begin{cases} B(y_o^2, \varepsilon_2/2) \ldots, B(y_{\ell_2}^2, \varepsilon_2/2) \\ y_o^2, \ldots, y_{\ell_2}^2 \in A(x_n) \end{cases}$$

On prend k_2 tel que pour tout $i \in \{0, 1, \ldots, \ell_2\}$, il existe $n \in \{\alpha(2), \ldots, k_2\}$ tel que $x_n \in B(y_i^2, \varepsilon_2/2)$. On définit alors $\beta(2) = \max\{k_2, \alpha(2), n(\varepsilon_3)\}$, $\alpha(3) = \beta(2)+1$ et on montre que :

$$\delta(X_{\alpha(2)}^{\beta(2)}, A(x_n)) \leq \varepsilon_2$$

La construction se poursuit de façon analogue, et on obtient finalement :

$$\sum_{i\in\mathbb{N}} \delta(X_{\alpha(i)}^{\beta(i)}, X_{\alpha(i+1)}^{\beta(i+1)}) \leq 2 \sum_{i\in\mathbb{N}} \delta(X_{\alpha(i)}^{\beta(i)}, A(x_n)) < \infty$$

□

Remarque. 1°) En fait la démonstration de la proposition établit le résultat plus précis suivant :

$$\begin{cases} \text{si } (x_n) \text{ est bornée et si } (\varepsilon_n)_{n\geq 1} \text{ est une suite de réels} > 0, \\ \text{alors il existe } \alpha(n) \text{ tel que } \forall m \geq 1 : \delta(X_{\alpha(m)}^{\alpha(m+1)-1}, A(x_n)) \leq \varepsilon_m \end{cases}$$

Ceci signifie que l'ensemble d'accumulation de (x_n) peut être approché aussi finement que l'on veut avec des tranches complémentaires de la suite (x_n). Ces résultats sont à rapporcher de certains résultats de [2].

2°) Si, à la place de (σ), on suppose seulement :

$$(\sigma') \lim_{i\to+\infty} \delta(X_{\alpha(i)}^{\beta(i)}, X_{\alpha(i+1)}^{\beta(i+1)}) = 0$$

ou :

$$(\sigma'') \lim_{i\to\infty} X_{\alpha(i)}^{\beta(i)} = A(x_n),$$

le théorème 1 ne reste plus vrai. C'est ce qu'établit le contre-exemple suivant :

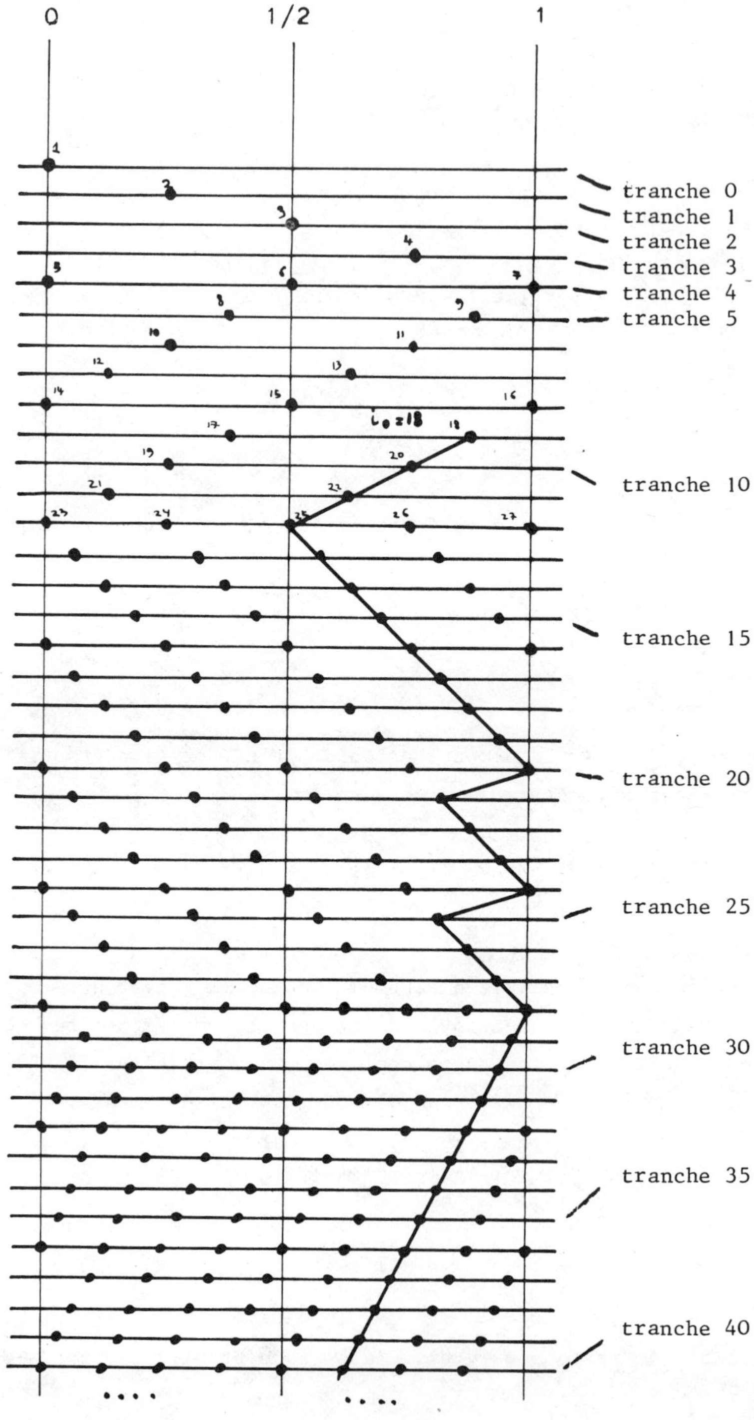
0
1/2
1
tranche 0
tranche 1
tranche 2
tranche 3
tranche 4
tranche 5
tranche 10
tranche 15
tranche 20
tranche 25
tranche 30
tranche 35
tranche 40
$i_0 = 18$

Figure 6.

Contre-exemple. Considérons la suite (x_n) représentée à la figure 6 :

$$(x_n) = (0, 1/4, 1/2, 3/4, 0, 1/2, 1, 3/8, 7/8, 1/4, 3/4, \ldots)$$

Prenons $\alpha(n) = (0, 1, 2, 3, 4, 7, 9, 11, 13, \ldots)$ (voir figure 6)

$\beta(n) = \alpha(n+1) - 1$

On a $A(x_n) = [0,1]$

Les conditions (σ') et (σ'') sont vérifiées mais pas la condition (σ). Il est clair que pour tout $i_o \in \mathbb{N}$, la suite (x_{i_j}) est une suite non convergente vérifiant : $A(x_{i_j}) = [0,1]$.

Théorème 2. Si (x_n) est bornée et n'a qu'un nombre fini de points d'accumulation et si l'une des deux conditions suivantes est satisfaite :

$$(\sigma') \lim_{i\to\infty} (X^{\beta(i)}_{\alpha(i)}, X^{\beta(i+1)}_{\alpha(i+1)}) = 0$$

$$(\sigma'') \lim_{i\to\infty} X^{\beta(i)}_{\alpha(i)} = A(x_n)$$

alors : (i) pour tout $i_o \in \mathbb{N}$: $i_o \downarrow$

(ii) pour tout $\hat{x} \in A(x_n)$, il existe $i_o \in \mathbb{N}$ tel que $i_o \downarrow \hat{x}$.

Remarque. $(\sigma) \Rightarrow (\sigma')$; $(\sigma) \Rightarrow (\sigma'')$. Les hypothèses (σ') et (σ'') sont donc plus faciles à satisfaire que l'hypothèse (σ). Il résulte aussi de la proposition 1 que (x_n) étant donnée, il est toujours possible de trouver $(\sigma(n), \beta(n))$ satisfaisant (σ'), (σ'').

Démonstration. Notons $y_1, y_2, \ldots, y_\ell$ les points d'accumulation de (x_n). Posons $\varepsilon_o = \min \{d(y_i, y_j) ; i \neq j\}$.

Montrons d'abord que puisque (x_n) est bornée et n'a qu'un nombre fini de points d'accumulation :

$$(\sigma') \Leftrightarrow (\sigma'')$$

a) $(\sigma') \Rightarrow (\sigma'')$. Soit i_o tel que :

$$\begin{cases} \forall i \geq i_o : \delta(X^{\beta(i)}_{\alpha(i)}, X^{\beta(i+1)}_{\alpha(i+1)}) \leq \varepsilon_o/4 \\ \forall m \geq \alpha(i_o) : d(x_m, A(x_n)) \leq \varepsilon_o/4 \end{cases}$$

En raisonnant par l'absurde, on établit que :

$$\forall i \geq i_o, \forall j \in \{1, 2, \ldots, \ell\} : B(y_j, \varepsilon/4) \cap X_{\alpha(i)}^{\beta(i)} \neq \emptyset.$$

Ceci permet alors de montrer que :

$$\liminf_{i\to\infty} X_{\alpha(i)}^{\beta(i)} \supset A(x_n)$$

Comme l'inclusion :

$$\limsup_{i\to\infty} X_{\alpha(i)}^{\beta(i)} \subset A(x_n)$$

est toujours vérifiée, on peut conclure.

b) (σ'') => (σ'). Pour tout $r \in \{1, \ldots, \ell\}$, soit $(x_{k_r(i)})$ une sous-suite de (x_n) convergeant vers y_r avec $k_r(i) \in \{\alpha(i), \ldots, \beta(i)\}$. Soit $\varepsilon \in \mathbb{R}^{+*}$ donné il existe $n(\varepsilon)$ tel que :

$$\begin{cases} \forall m \geq n(\varepsilon) : d(x_m, A(x_n)) \leq \varepsilon \\ \forall r \in \{1,\ldots,\ell\} : k_r(m) \geq n(\varepsilon) \Rightarrow d(x_{k_r(m)}, y_i) \leq \varepsilon \end{cases}$$

Il est clair alors que pour tout i tel que $\alpha(i) \geq n(\varepsilon)$ on a :

$$\delta(X_{\alpha(i)}^{\beta(i)}, X_{\alpha(i+1)}^{\beta(i+1)}) \leq 2\varepsilon$$

ce qui démontre donc (σ').

c) Partie (i) du théorème 2 : un raisonnement semblable à celui fait pour le théorème 1 montre que quel que soit $i_o \in \mathbb{N}$:

$$\lim_{i\to\infty} d(t_i, t_{i+1}) = 0.$$

Il en résulte que $A(t_i)$ est connexe [2] [7]. D'autre part, $A(t_i) \subset A(x_n)$. Donc (t_i) converge.

d) Partie (ii) du théorème 2 : soit y_r un point d'accumulation de (x_n) On prend i_o de la façon suivante :

$$\begin{cases} \alpha(i) \geq i_o \Rightarrow \delta(X_{\alpha(i)}^{\beta(i)}, X_{\alpha(i+1)}^{\beta(i+1)}) \leq \varepsilon_o/4 \\ m \geq i_o \Rightarrow d(x_m, A(x_n)) \leq \varepsilon_o/4 \\ d(x_{i_o}, y_r) \leq \varepsilon_o/4. \end{cases}$$

□

Théorème 3. (i) Si (x_n) est pseudo-périodique, alors en prenant : $\alpha(n) = n$ $\beta(n) = 2n$, on a :

$$\forall i_o : i_o \downarrow$$

$$\forall \hat{x} \in A(x_n), \exists i_o : i_o \downarrow \hat{x}.$$

(ii) Si $\hat{x}$ est un point d'accumulation isolé de (x_n) et ayant une force strictement positive, alors en prenant : $\alpha(n) = n$, $\beta(n) = n^2$, on a :

$$\exists i_o : i_o \downarrow \hat{x}$$

Rappelons que l'on dit que (x_n) est pseudo-périodique (ou asymptotiquement cyclique) s'il existe $p \geq 1$ tel que les sous-suites (x_{np}), (x_{np+1}),.. .., (x_{np+p-1}) soient convergentes.

Démonstration. (i) Les conditions du théorème 2 sont satisfaites.

(ii) Soit $\varepsilon = d(\hat{x}, A(x_n) - \{\hat{x}\}) > 0$.

Il existe n_o tel que si $n \geq n_o$ alors :

et $$\begin{cases} X_n^{n^2} \cap B(\hat{x}, \varepsilon/5) \neq \emptyset \\ x_n \in B(\hat{x}, \varepsilon/5) \text{ ou } x_n \notin B(\hat{x}, 3\varepsilon/5). \end{cases}$$

On prend alors : $i_o \geq n_o$ tel que $d(x_{i_o}, \hat{x}) \leq \varepsilon/5$.

□

IV. DESCRIPTION DES ALGORITHMES S ET EXEMPLES

On se donne (x_n), $(\alpha(n), \beta(n))$ comme dans II. Soit $a \in E$.

ALGORITHME $S(a, \alpha(n), \beta(n))$.

Etape j. Soit I_j l'ensemble des entiers ℓ tels que :

$$\begin{cases} \alpha(j) \le \ell \le \beta(j) \\ d(x_\ell, a) = \min \{d(x_m, a) \mid \alpha(j) \le m \le \beta(j)\} \end{cases}$$

Soit $i_j = \max I_j$. On pose : $s_j = x_{i_j}$.

Une fois les paramètres a, $(\alpha(n), \beta(n))$ fixés, l'algorithme S extrait une sous-suite (s_j) en prenant, dans chacune des tranches définies par $\alpha(n)$, $\beta(n)$, le point le plus proche de a (lorsqu'il y a plusieurs points réalisant le minimum, il choisit celui de plus grand indice).
Si la sous-suite $(x_{i_j}) = (s_j)$ converge vers $\hat{x}$, on écrira :

$$(a, \alpha(n), \beta(n)) \uparrow \hat{x}$$

ou même $a \uparrow \hat{x}$.

Pour exprimer que $(x_{i_j}) = (s_j)$ converge, on écrira :

$$(a, \alpha(n), \beta(n)) \uparrow$$

ou même $a \uparrow$.

Exemple 1. Prenons $E = \mathbb{R}$ et considérons (x_n) définie par :

$$x_{2n} = -(1 + 1/2^n)$$

$$x_{2n+1} = +(1 + 1/2^n).$$

Fixons dans un premier temps : $\alpha(n) = 2n$, $\beta(n) = 2n+1$

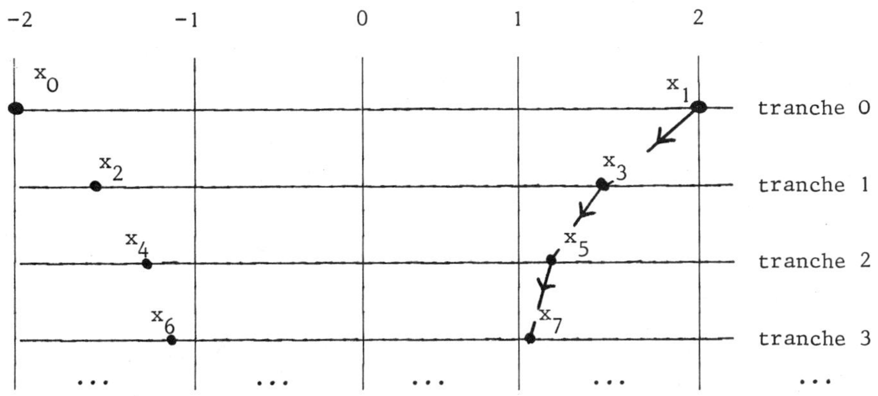

Figure 7.

Si on prend $a \geq 0$, on obtient : $a\uparrow \ +1$.

Si on prend $a < 0$, on obtient : $a\uparrow \ -1$.

On a donc : (*) $\forall a : a\uparrow$

(**) $\forall \hat{x} \in A(x_n) \quad \exists a : a \uparrow \hat{x}$.

Fixons maintenant

$$\alpha(n) = 3n, \ \beta(n) = 3n+2.$$

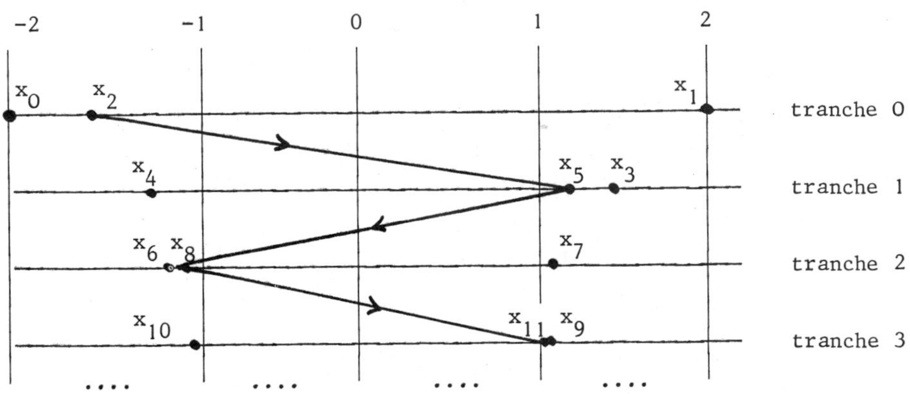

Figure 8.

Si on prend : $a > 0$, on obtient : $a \uparrow +1$.

Si on prend : $a < 0$, on obtient : $a \uparrow -1$.

Si on prend : a = 0, on obtient : $(s_i) = (x_2, x_5, x_8, x_{11}, \ldots)$ et donc (s_i) ne converge pas.

Exemple 2. $E = \mathbb{C}$

$$(x_n) = e^{in}(1+1/2^n)$$
$$A(x_n) = \{z \in \mathbb{C} \mid |z| = 1\}$$

Si on choisit $\beta(n) - \alpha(n) < C$, alors aucun point a ne fournira de suite convergente.

Si on choisit : $\alpha(n) = 2^n$

$$\beta(n) = \alpha(n+1) - 1$$

alors avec : arg a = θ, on aura : $a \uparrow e^{i\theta}$ et avec : a = 0, on aura : $A(s_i) = A(x_n)$

V. THEOREME DE CONVERGENCE POUR S

Théorème. Si les conditions suivantes sont vérifiées :

$$(\sigma'') \quad \lim_{i \in \mathbb{N}} X_{\alpha(i)}^{\beta(i)} = A(x_n)$$

$$(\gamma) \quad \text{card } \{x \in A(x_n) \mid d(a, x) = d(a, A(x_n))\} = 1$$

alors $a\uparrow$.

Remarque. D'après la proposition 1 (voir aussi [2]), il est toujours possible de choisir $(\alpha(n), \beta(n))$ pour que (σ'') soit vérifiée. De même il est clair qu'il est toujours possible de prendre a vérifiant (γ). Les conditions de ce théorème sont malgré tout moins commodes à utiliser que celles du théorème 1, car elles font intervenir de façon essentielle $A(x_n)$.

Démonstration. Soit $\hat{x}$ le seul élément de $A(x_n)$ tel que :

$$d(a, \hat{x}) = d(a, A(x_n)).$$

Montrons que : $\lim_{j\to\infty} s_j = \hat{x}$. Par hypothèse : $\hat{x} \in \liminf_{j\to\infty} X_{\alpha(j)}^{\beta(j)}$. Il existe donc une sous-suite (x_{k_j}) convergente vers $\hat{x}$ et telle que :

$$\forall j \in \mathbb{N} : x_{k_j} \in X_{\alpha(j)}^{\beta(j)}.$$

Donc :

$$\lim_{j\to\infty} d(a, x_{i_j}) = d(a, \hat{x}).$$

Cette dernière relation avec le fait que les boules fermées de E sont compactes, implique que :

$$\lim_{j\to\infty} x_{i_j} = \hat{x}.$$

□

VI. IMPOSSIBILITE D'UN ALGORITHME D'EXTRACTION TOUJOURS CONVERGENT.

Pour terminer nous allons établir qu'un algorithme réussissant à extraire une sous-suite convergente de n'importe quelle suite non convergente ne peut pas exister.

Soit F une transformation de suites : $F : \mathcal{S}(E) \to \mathcal{S}(E)$

$: (x_n) \to F(x_n) = (t_n).$

On dit que F est une transformation pour l'extraction si (t_n) est toujours une sous-suite de (x_n). On dit que F est un algorithme d'extraction si, de plus, F peut être définie par un algorithme pour suites [2] [4], ce qui signifie que chaque t_i ne dépend que d'un nombre fini de points de la suite (x_n) (ce nombre fini pouvant varier avec i, et (x_n)).

Par exemple, les transformations :

$$F : (x_n) \to (x_{2n}) \; ; \; F : (x_n) \to (\max(x_{2n+1}, x_{2n}))$$

sont des algorithmes d'extraction.

Lorsque i_o, $(\alpha(n), \beta(n))$ sont fixés $T(i_o, \alpha(n), \beta(n))$ est <u>un</u> algorithme d'extraction. De même lorsque a, $(\alpha(n), \beta(n))$ sont fixés $S(a, \alpha(n), \beta(n))$ est <u>un</u> algorithme d'extraction.

<u>Théorème</u>. Si E possède plus de deux points alors il n'existe aucun algorithme d'extraction F tel que toute suite $(x_n) \in \mathcal{S}(E)$ donne par F une sous-suite (t_n) convergente.

<u>Remarques</u>. 1) Comme le montre la démonstration, le théorème reste vrai en remplaçant "toute suite $(x_n) \in S(E)$" par "toute suite bornée (x_n)" ou même par "toute suite (x_n) n'ayant qu'un nombre fini de points d'accumulation".

2) Ce résultat montre en particulier qu'aucun choix de i_o, $\alpha(n)$, $\beta(n)$ ne pourra être satisfaisant pour toute suite (x_n) (chose qui était intuitivement assez claire), mais il montre aussi qu'aucun algorithme (aussi compliqué soit-il, et même basé sur des principes totalement différents de ceux étudiés dans les paragraphes précédents) ne pourra extraire une sous-suite convergente de toute suite non convergente.

3) Ce théorème est à rapprocher d'autres résultats obtenus à propos de l'accélération de la convergence [6].

Démonstration. Soient a, b $\in$ E, a $\neq$ b. Raisonnons par l'absurde en supposant donné F un algorithme d'extraction tel que pour toute suite $(x_n) \in S$, la suite extraite par F de (x_n) soit convergente.
Soit $(x_n^o) = (a, a, a, \ldots)$. Notons (t_n^o) la suite obtenue par F avec (x_n^o).
On a nécessairement : $t_n^o = a$ pour tout n.
Posons m(0) = 0. Soit n(0) le plus grand indice des points de (x_n^o) ayant servi au calcul de $t_{m(o)}^o$.
Soit la suite :

$$(x_n^1) = (x_o^o, x_1^o, \ldots, x_{n(o)}^o, b, b, \ldots).$$

Notons (t_n^1) la suite obtenue par F avec (x_n^1). On a nécessairement : $t_{m(o)}^1 = a$ (car (x_m^1) commence par les mêmes (n(0)+1) premiers points que (x_n^o)).
D'autre part (par hypothèse sur F), il existe m(1) > m(0) tel que :

$$t_{m(1)}^1 = b.$$

Soit n(1) le plus grand indice des points de (x_n^1) ayant servi au calcul des t_i^1, $i \in \{0, 1, \ldots, m(1)\}$.
Soit la suite :

$$(x_n^2) = (x_o^1, x_1^1, \ldots, x_{n(1)}^1, a, a, \ldots).$$

Notons (t_n^2) la suite obtenue par F avec (x_n^2). On a nécessairement :

$$t_{m(o)}^2 = a, \; t_{m(1)}^2 = b.$$

Il existe m(2) > m(1) tel que $t_{m(2)}^2 = a$. etc ...
Considérons la suite :

$$(x_n) = (x_o^o, x_1^o, \ldots, x_{n(o)}^o, x_{n(o)+1}^1, \ldots, x_{n(1)}^1, x_{n(1)+1}^2, \ldots, x_{n(2)}^2, \ldots).$$

Notons (t_n) la suite obtenue par F avec (x_n). Par construction :

$$\begin{cases} t_{m(i)} = a, \text{ si } i \text{ pair}, \\ t_{m(i)} = b, \text{ si } i \text{ impair}. \end{cases}$$

(t_n) n'est donc pas convergente, ce qui contredit l'hypothèse faite sur F.

REFERENCES

1. C. BERGE, Espaces topologiques- Fonctions multivoques. Dunod, Paris 1966.

2. J.P. DELAHAYE, Quelques problèmes posés par les suites de points non convergentes et algorithmes pour traiter de telles suites. Thèse de 3e cycle, Lille 1979.

3. J.P. DELAHAYE, Expériences numériques sur les algorithmes d'extraction pour suites non convergentes. Publication ANO n° 5, Université des Sciences et Techniques de Lille, 1979.

4. J.P. DELAHAYE, Algorithmes pour suites non convergentes, Numerische Math., 34, 1980, pp 333-347.

5. J.P. DELAHAYE et J. DENEL
The continuities of point-to-set maps ; definitions and equivalence. Mathematical Programming Study 10, 1979, p 8-12.

6. J.P. DELAHAYE et B. GERMAIN-BONNE, Résultats négatifs en accélération de la convergence. Numerische Math., 35, 1980, pp 443-457.

7. A.M. OSTROWSKI, Solution of equations and systems of equations. Academic Press, New York 1966.

THE n-STEP SQUARE CONVERGENCE OF SOME MINIMIZATION ALGORITHMS RELATED TO POWELL's DERIVATIVE FREE METHOD

Josef Stoer

University of Würzburg
Würzburg, West Germany

I. INTRODUCTION

This paper is concerned with the convergence of certain algorithms for solving unconstrained minimization problems of the form

$$\text{(I.1)} \qquad \text{Find} \quad \min_{x \in R^n} \quad f(x)$$

where $f: R^n \to R$ is a given real valued function on R^n. In particular we are interested in their local convergence within a neighborhood $U(\bar{x})$ of a strict local minimum of f where f satisfies the following assumptions:

(A) 1) $f \in C^2(U(\bar{x}))$: f is twice continuously differentiable for $x \in U(\bar{x})$.

2) The gradient $g(x)$ of f vanishes at $\bar{x}: g(\bar{x}) = 0$.

3) There are constants $m, M > 0$ such that the hessian $h(x)$ of f satisfies

$$m\|y\|^2 \leq y^T h(\bar{x})y \leq M\|y\|^2$$

for all $y \in R^n$.

4) $h(x)$ is Lipschitz continuous at $\bar{x}$: There is a constant Λ such that $\|h(x) - h(\bar{x})\| \leq \Lambda \cdot \|x-\bar{x}\|$ for all $x \in U(\bar{x})$.

Here, $\|\cdot\|$ denotes the euclidean norm or its subordinate matrix norm, respectively. Almost all known algorithms for solving (I.1) are composed of line searches. They generate a sequence $\{x_i\}$ of approximations to $\bar{x}$ such that

$$x_{i+1} = x_i - \lambda_i s_i \tag{I.2}$$

where s_i is some search direction and the step-size λ_i is determined by a line search in such a way that

$$f(x_{i+1}) \approx \min_{\lambda} f(x_i - \lambda s_i)$$

holds at least approximately. The choice of s_i depends on the particular algorithm but usually s_i is computed from the gradient $g_i = g(x_i)$ and other auxiliary parameters a_i , say $a_i \in R^n$, according to some rule typical for the algorithm,

$$s_i = \varphi(g_i, a_i) \ . \tag{I.3}$$

Also, within most methods the parameter a_i is updated after having computed x_{i+1} by means of some up-date formula of the form

$$a_{i+1} = \Psi_i(p_i, g_i, g_{i+1}, a_i) \ , \ p_i := x_{i+1} - x_i \ . \tag{I.4}$$

In many cases, the parameters a_i and functions φ and Ψ_i have been devised in such a way, that the corresponding iteration method terminates with the exact solution of (I.1) within at most n iterations for all starting points $x_o \in R^n$ and all suitably chosen starting parameters a_o, if it is applied to strictly convex quadratic functions

$$f(x) = \frac{1}{2} x^T Cx - b^T x + c \quad ,$$

with C a positive definite $n \times n$-matrix, provided exact line searches are used in every step ("n-step quadratic termination property").
A systematic convergence theory for methods of this type has been described in [8,I,II], where it is shown that in general under some weak conditions on the functions φ, Ψ_i, the n-step quadratic termination property implies the n-step square convergence of the method, $\|x_{i_k+n}-\bar{x}\| \leqslant c \|x_{i_k}-\bar{x}\|^2$, $k \geqslant 0$, for certain subsequences of $\{x_i\}$. Whereas this theory can be applied to almost all minimization methods having the n-step quadratic termination property (e.g. the conjugate gradient algorithms of Fletcher-Reeves, the DFP, BFGS variable metric algorithms, and more generally most algorithms of Broyden's β-class and the still more general Oren-Luenberger class of rank-2 algorithms), there are still some important methods with certain quadratic termination properties which cannot be easily covered by the theory of [8,I,II]. These methods are all related to the derivative-free method of Powell [5] and have been studied by several authors (Brent [1], Chazan, Miranker [2], Hestenes [3], Sloboda [6], Zangwill [13]). These methods are also composed of line searches

$$x_i \to x_{i+1} = x_i - \lambda_i s_i \approx \arg\min_{\lambda} f(x_i - \lambda s_i) \; ,$$

but unlike the methods considered in [8,I,II], the search direction s_i cannot be easily described by simple formulae of the type (I.3),(I.4). Nevertheless, essentially the same techniques as in [8,I,II], can be applied to investigate their convergence properties and it is the purpose of this paper to present the corresponding results.
Section II summarizes some known simple technical lemmas for an easy reference in subsequent sections.
Section III deals with the convergence of a variant of Powell's derivative-free method [5], which also covers

Brent's algorithm [1] and is related but not identical to the Zangwill's [13] variant of Powell's method. A similar variant has been investigated by Sutti [9,10] and shown to be convergent, however no proof of its n-step square convergence was given so far. Toint and Callier [11,12] proved some results on the Q-superlinear convergence of Powell's method, but only under some restrictive hypotheses, which seem not to hold in general and are hard to verify. Whereas the methods of section 3 are sequential methods, the algorithms investigated in section 4 due to Chazan, Miranker [2] and Sloboda [6] (see also Hestenes [3]) are parallel methods. The main results of the paper are contained in theorems(III.35) and (IV.16)which essentially assert the n-step square convergence of these methods.

II. SOME AUXILIARY RESULTS.

For future reference, we first list some easy consequences of hypothesis (A) without proof. In order to simplify the notation we may assume $\bar{x} = 0$ and $f(0) = 0$ without loss of generality. Denote by $C := h(0)$ the hessian of f at 0, so that $\tilde{f}(x) = \frac{1}{2}x^T Cx$ is a quadratic approximation to f with gradient $\tilde{g}(x) := Cx$ and hessian $\tilde{h}(x) \equiv C$. Then the following holds under assumption (A):

<u>There are numbers</u> $R, m, M, \Lambda > 0$ <u>such that</u>

a) $$|f(x) - \tilde{f}(x)| < \frac{\Lambda}{6} \|x\|^3$$

b) $$\|g(x) - \tilde{g}(x)\| \leq \frac{\Lambda}{2} \|x\|^2$$

(II.1)

c) $$\|h(x) - \tilde{h}(x)\| \; \Lambda \; \|x\|$$

d) $$\frac{m}{2}\|y\|^2 \leq y^T h(x) y \leq 2M\|y\|^2$$

<u>for all</u> $y \in R^n$, $x \in K_R := \{z \mid \|z\| \leq R\}$.

Note that $\tilde{f}(x) = \frac{1}{2} x^T Cx$ is a strictly convex quadratic function with C a positive definite $n \times n$-matrix satisfying

(II.1') $$m\|x\|^2 \leq x^T Cx \leq M\|x\|^2 .$$

As a consequence,

(II.2) f <u>is strictly convex on</u> K_R <u>and has</u> 0 <u>as a strict local minimum</u>.

Hence the restriction $L_R(x) := \{z | f(z) \leq f(x)\} \cap K_R$ of the usual level set $\{x | f(z) \leq f(x)\}$ to K_R is convex and bounded for all $x \in K_R$. This result can be strengthened in the following way (see for instance [8,I] for a proof).

Denote by $L(x)$ the connected component of the level set $\{z | f(z) \leq f(x)\}$ that contains x. Then, under assumption (II.1),

(II.3) <u>there is a number</u> ρ, $0 < \rho < R$ <u>such that</u> $L(x)$ <u>is a compact convex set with</u> $0 \in L(x) \subset K_R$ <u>for all</u> $\|x\| \leq \rho$.

In order to describe the line search, we assume that λ_i is always chosen such that

(II.4) $$x_{i+1} = x_i - \lambda_i s_i \in L(x_i) ,$$

which ensures $f(x_{i+1}) \leq f(x_i)$. As a measure of the inexactness of the line search we use the number

(II.5) $$\mu_i := \begin{cases} g_{i+1}^T s_i / g_i^T s_i , & \text{if } g_i^T s_i \neq 0 \\ 0 , & \text{otherwise,} \end{cases}$$

Note that for $\|x_i\| < \rho$ we have $\mu_i = 0$ if and only if

$$f(x_{i+1}) = \min\{f(x_i - \lambda s_i) | x_i - \lambda s_i \in L(x_i)\} .$$

We say that the line search is <u>asymptotically perfect</u>, if there is a constant $\alpha > 0$ such that $|\mu_i| \leq \alpha \cdot \|g_i\|$ for all i .

As a simple consequence of $f(x_j) \leqslant f(x_i)$, $j \geqslant i$, we note for $\|x_i\| \leqslant \rho$ the inequality

$$\text{(II.6)} \qquad \|x_j\| \leqslant 2\sqrt{\frac{M}{m}}\,\|x_i\| \quad \text{for} \quad j > i\,.$$

This follows from (II.1)

$$\frac{m}{4}\|x_j\|^2 \leqslant f(x_j) \leqslant f(x_i) \leqslant M\|x_i\|^2 \quad .$$

In the following we assume that $|\mu_i| \leqslant \bar{\mu} < 1$ for all i. Then for small $\|x_o\|$, the line searches used are completely specified by the sequence $\{\mu_i\}$. This results from (see [8,I]

(II.7) Under assumption (2.1), for each number $0 \leqslant \bar{\mu} < 1$ there is an $\epsilon > 0$ such that for all $\|x\| \leqslant \epsilon$, $|\mu| \leqslant \bar{\mu}$, $\|s\| = 1$, there is a unique $x' = x - \lambda s \in L(x)$ with $g(x')^T s = \mu \cdot g^T s$.

(II.7) says that there is a line search function

$$e(x,s,\mu,g) := x' = x - \lambda s\,,$$

which is well defined for $\|x\| \leqslant \epsilon$, $|\mu| \leqslant \bar{\mu}$, $\|s\| = 1$ and $g(.)$ satisfying (A), (II.1). The following important proposition (see [8,I] for a proof) investigates the effect of changes in the arguments of $e(.)$:

(II.8) PROPOSITION. Let f satisfy (A), (II.1) and let $\tilde{g}(x) := Cx$, $C = h(0)$ as in (II.1). Then for all $0 \leqslant \alpha$ there are constants $\epsilon > 0$, σ_k which only depend on m, M, Λ and α, such that

$$\|e(x,s,\mu,g) - e(\tilde{x},\tilde{s},0,\tilde{g})\| \leqslant \sigma_1\|\Delta x\| + \sigma_2\|\tilde{x}\|\|\Delta s\| + \sigma_3\|\Delta g\| + \sigma_4\|x\|^2 + \sigma_5\|x\|\|\tilde{x}\|$$

for all $\|x\|$, $\|\tilde{x}\| \leqslant \epsilon$, all μ with $|\mu| \leqslant \alpha\|g(x)\|$ and all s, $\tilde{s}$ with $\|s\| = \|\tilde{s}\| = 1$. Here Δx, Δs, and Δg mean $\Delta x = x - \tilde{x}$, $\Delta s = s - \tilde{s}$ and $\Delta g = g(x) - \tilde{g}(\tilde{x})$, respectively.

III. THE CONVERGENCE OF POWELL's DERIVATIVE-FREE METHOD

In this section we analyze the convergence of Powell's derivative-free minimization method [5]. A numerically stable reformulation of this method has been given by Brent [1]; also, Kaufman [4] recently used this method as a basis for solving linearly constrained minimization problems; see also Sutti [9,10]. As the original version of this method was ill defined in certain situations, we consider the following slightly different variant, which also covers Brent's formulation. We only describe its basic cycle, leading from an initial approximation x_o for the minimum of a function $f: R^n \to R$ to better approximations x_j, $j=0,1,\ldots,j_o$ for some $j_o \leq n+1$, which might be further improved using the same cycle with x_{j_o} as new starting value, etc.

(III.1) Given $x_o \in R^n$, do:

1) Choose a suitable orthogonal $n\times n$-matrix Ω and $D_o := \Omega$.

2) For $j := 0,1,\ldots, n$ do:

a) Set $y_o^{(j)} := x_j$, $(d_o^{(j)}, d_1^{(j)},\ldots,d_{n-1}^{(j)}) := D_j$

b) For $k := 0,1,\ldots,n-1$ compute $\lambda_k = \lambda_k^{(j)}$ to minimize $f(y_k^{(j)} - \lambda_k d_k^{(j)})$ approximately and set $y_{k+1}^{(j)} := y_k^{(j)} - \lambda_k^{(j)} d_k^{(j)}$.

c) For $j < n$ define $i = i(j)$ as the smallest $i \geq 0$ with $|\lambda_i^{(j)}| := \max\{|\lambda_k^{(j)}| :$

$: 0 \leq k \leq n-j-1\}$

d) If $j < n$ and $\lambda_i^{(j)} \neq 0$ then

α) set $d^{(j)} := p/\|p\|$, where $p := y_o^{(j)} - y_n^{(j)}$

$$D_{j+1} := (d_o^{(j)}, \ldots, d_{i-1}^{(j)}, d_{i+1}^{(j)}, \ldots, d_{n-1}^{(j)}, d^{(j)}) ,$$

β) <u>compute</u> $\lambda = \lambda^{(j)}$ <u>to minimize</u> $f(x_j - \lambda d^{(j)})$ <u>approximately, set</u>

$$x_{j+1} := x_j - \lambda^{(j)} d^{(j)} ,$$

<u>increase</u> j <u>by</u> 1 <u>and go to</u> 2a);

<u>otherwise</u>

a) α) <u>if</u> $\lambda_k^{(j)} = 0$ <u>for</u> $k=0,1,\ldots,n-1$ <u>then</u> <u>set</u> $j_o := j$ <u>and stop</u>: x_{j_o} <u>is optimal</u> ;

β) <u>if</u> $\lambda_k^{(j)} \neq 0$ <u>for some</u> k , <u>set</u> $j_o := j+1$, $x_{j_o} := y_n^{(j)}$ <u>and stop</u> .

As before, we say that the approximate minimizations 2b), 2dβ) of (III.1) are asymptotically perfect if there is a constant α , such that

$$g_{k+1}^{(j)\,T} d_k^{(j)} = \mu_k^{(j)} g_k^{(j)\,T} d_k^{(j)}, \quad |\mu_k^{(j)}| \leq \alpha \cdot \|g_k^{(j)}\|$$

(III.2)

$$g_{j+1}^T d^{(j)} = \mu_j g_j^T d^{(j)} \qquad |\mu_j| \leq \alpha \|g_j\| ,$$

for all j and all $k = 0,1,\ldots,n-1$ where $g_k^{(j)} := g(y_k^{(j)})$, $g_j := g(x_j)$. Of course, (III.2) cannot be checked directly without computing the gradients $g_k^{(j)}$, g_j and so there is a certain problem to decide whether any given mode of line search satisfies (III.2) or not. We will not pursue this question.

Note that (III.1) is almost a type-II algorithm in the sense of [8,I,II], if we identify the parameters a_j with the matrices D_j: It is composed of line searches and there is an update formula for the parameters D_i . The only difficulty is that the update functions $D_j \to D_{j+1}$ have not the form (I.4): If we identify the iterates x_j of (I.2) with the x_j of (III.1), then the update D_{j+1} of D_j is computed by means of the intermediate line searches

$$x_j = y_o^{(j)} \to y_1^{(j)} \to \dots \to y_n^{(j)} \to x_{j+1} \quad ,$$

which cannot be described by a function Ψ of the form (I.4). If we identify the iterates x_j of (I.2) with all iterates of (III.1), then again Ψ cannot be described by (I.4), since by (III.1), 2d)α), D_{i+1} is computed from $y_o^{(j)} - y_n^{(j)}$ rather than from the two latest iterates $y_{n-1}^{(j)} - y_n^{(j)}$, as is supposed in (I.4). There is the further technical difficulty, that the formula for D_{j+1} involves a pivot selection (step 2c) of (III.1), which makes the update function discontinuous for certain arguments. Nevertheless, the theory of [8,I,II] carries over to the present case with only minor changes.

We first show some properties of (III.1) when applied to quadratic functions (compare Powell [5]):

(III.3) THEOREM: Let $\tilde{f}(x) = \frac{1}{2} x^T Cx$ be a strictly convex quadratic function on R^n with (II.1') and suppose (III.1) is applied to $\tilde{f}$ with the following modifications:

i) all line searches are exact;

ii) step 2c) is weakened to:

(III.4)2c') For $j < n$ find any i with $0 \leqslant i \leqslant n-j-1$ and $\lambda_i^{(j)} \neq 0$. If there is none, set $i := 0$.

Then for all $\tilde{x}_o \in R^n$ and all orthogonal starting matrices $\Omega := \tilde{D}_o$, (III.1) generates iterates $\tilde{x}_j, \tilde{D}_j$, $j \geqslant 0$, with the following properties:

a) There is a first index $l = j_o$, $0 \leqslant l \leqslant n$ such that $\tilde{x}_l = 0$ and the algorithm stops via step 2eα) of (III.1) with the optimal solution (n-step square convergence).

b) *For all* $0 \leqslant j \leqslant 1$ *the matrices*

$$\tilde{D}_j = (\tilde{d}_o^{(j)}, \ldots, \tilde{d}_{n-1}^{(j)}$$

are well-defined nonsingular $n \times n$-*matrices. Its last* j *columns are* C-*conjugate*:

$$\tilde{d}_\mu^{(j)T} C\tilde{d}_\nu^{(j)} = 0 \quad \text{for} \quad n-j \leqslant \mu < \nu \leqslant n-1 \; .$$

The *proof* is by induction on j. For $j = 0$, (III.3)b) is trivial. Assume (III.3)b) and $\tilde{x}_j \neq 0$ are true for some fixed $j \geqslant 0$. Use the simpler notation $y_k = \tilde{y}_k^{(j)}$, $d_k := \tilde{d}_k^{(j)}$, $\lambda_k := \tilde{\lambda}_k^{(j)}$, $g_k := Cy_k$. Then, for the index j, (III.1)2) generates y_k, $k=0,1,\ldots,n$, with the following properties: By induction hypothesis, y_o and y_n are both the result of minimizing $\tilde{f}$ exactly into the j conjugate directions $d_{n-j},\ldots,d_{n-1}$:

$$\tilde{y}_{n-j+1}^{(j-1)} \to \tilde{y}_{n-j+2}^{(j-1)} \to \ldots \to \tilde{y}_n^{(j-1)} \to \tilde{x}_j = y_o = \tilde{y}_o^{(j)}$$

$$\tilde{y}_{n-j}^{(j)} = y_{n-j} \to y_{n-j+1} \to \ldots \to y_n = \tilde{y}_n^{(j)} \quad .$$

Hence, by a familiar property of quadratic functions, $y_o - y_n$ is C-conjugate to these directions:

$$\text{(III.5)} \quad d_k^T C(y_o - y_n) = d_k^T g_o - d_k^T g_n = 0 \text{ for } n-j \leqslant k \leqslant n-1 \; .$$

By definition of y_n,

$$\text{(III.6)} \quad y_o - y_n = \lambda_o d_o + \ldots + \lambda_{n-j-1} d_{n-j-1} + \lambda_{n-j} d_{n-j} + \ldots + \lambda_{n-1} d_{n-1} \, ,$$

where because of the exact line search

$$\lambda_k = d_k^T g_k / d_k^T C d_k \quad .$$

Suppose $\lambda_k = 0$ for all $0 \leqslant k \leqslant n-j-1$. Then by (III.6)

$$g_o - g_n = \lambda_{n-j}Cd_{n-j} + \dots + \lambda_{n-1}Cd_{n-1}$$

and (III.5) together with the induction hypothesis gives $\lambda_k = 0$ for $k \geqslant n-j$ as well, so that $y_o = y_1 = \dots = y_n$, $g_o = \dots = g_n$ and therefore

$$0 = \lambda_k d_k^T C d_k = d_k^T g_k = d_k^T g_o \quad \text{for all} \quad 0 \leqslant k \leqslant n-1 .$$

Hence $g_o = y_o = \tilde{x}_j = 0$, since $\tilde{D}_j$ is nonsingular. Hence, if $\tilde{x}_j \neq 0$, there is a k with $0 \leqslant k \leqslant n-j-1$ and $\lambda_k = \tilde{\lambda}_k^{(j)} \neq 0$ and $p = y_o - y_n \neq 0$ by (III.6) and the non-singularity of $\tilde{D}_j$. Hence $\tilde{\lambda}_i^{(j)} \neq 0$ and $\tilde{D}_{j+1}$ is non-singular, and by (III.5) the last $j+1$ columns of $\tilde{D}_{j+1}$ are C-conjugate. Also, we see from these arguments that if $\tilde{\lambda}_i^{(j)} = 0$ for some $j < n$ then $\tilde{\lambda}_k^{(j)} = 0$ for all $0 \leqslant k \leqslant n-1$ and the algorithm stops via 2eα) with the exact solution $\tilde{x}_j = 0$. If $\tilde{\lambda}_i^{(j)} \neq 0$ for some $j < n$, the algorithm continues. If $j = n$, then all columns of $\tilde{D}_j$ are C-conjugate and therefore $\tilde{x}_j = \tilde{x}_n$ has been obtained after minimizing into these n conjugate directions, hence $\tilde{x}_n = 0$ is optimal and $\tilde{\lambda}_k^{(n)} = 0$ for all $k=0,1,\dots,n-1$, so that (III.1) again stops via 2eα) with the exact solution. This completes the proof. □

Thus counting only the main iterations $x_j \to x_{j+1}$ as steps, the algorithm (III.1) has the n-step quadratic termination property. We now try to establish an analogoue of lemma (3,9) of [8,I]. For this purpose, we first derive the boundedness of certain normalized quantities generated by (III.1) for quadratic functions $\tilde{f}$. Let $\mathfrak{M}_o$ be any algorithm obtained from (III.1), by using only exact line searches and replacing the pivoting rule 2c) of (III.1) by

(III.7) 2c") <u>For</u> $j < n$ <u>choose any</u> $i = i(j)$ <u>with</u>

$$|\lambda_i^{(j)}| \geqslant \tfrac{1}{2} \max\{|\lambda_k^{(j)}| : 0 \leqslant k \leqslant n-j-1\} .$$

Note that theorem(III.3)applies to $\mathfrak{M}_o$; moreover the following holds:

(III.8) THEOREM: If $\mathfrak{M}_o$ is applied to a strictly convex quadratic function $\tilde{f}(x) = \frac{1}{2} x^T Cx$ on R^n with (II.1'), then for all starting values $\tilde{x}_o \in R^n$, $\tilde{D}_o := \Omega$, Ω any orthogonal $n\times n$-matrix, the finitely many (see (III.3)) iterates $\tilde{x}_j$, $\tilde{y}_k^{(j)}$, $\tilde{D}_j$, $j \leqslant l$, $k \leqslant n-1$ generated by $\mathfrak{M}_o$ have the following properties: For all $0 \leqslant j < l$

a) $\|\tilde{D}_{j+1}\| \leqslant \beta$, $\|\tilde{D}_{j+1}^{-1}\| \leqslant \beta$

b) $\|\tilde{g}_o^{(j)}\| / \|\tilde{y}_o^{(j)} - \tilde{y}_n^{(j)}\| \leqslant \beta$, $\tilde{g}_o^{(j)} := C\tilde{y}_o^{(j)}$

c) $\max\{|\tilde{y}_k^{(j)}| : 0 \leqslant k \leqslant n-j-1\} \geqslant \|\tilde{g}_o^{(j)}\|/\beta$

where β is a constant which only depends on n, m and M , but not on $\tilde{x}_o$ and Ω .

Proof: Consider any fixed j with $0 \leqslant j < l$. Let us assume first that $\tilde{D}_j^{-1}$ exists. In order to simplify the notation, set

$$D := (d_o, d_1, \ldots, d_{n-1}) := \tilde{D}_j$$

$$\bar{D} := (\bar{d}_o, \bar{d}_1, \ldots, \bar{d}_{n-1}) := \tilde{D}_j \cdot \Delta_j^{-1}$$

with $\Delta_j := \mathrm{diag}(\delta_o, \ldots, \delta_{n-1})$, $\delta_k := \sqrt{d_k^T C d_k}$. Also, set

$$y_k := \tilde{y}_k^{(j)} , \ g_k := Cy_k , \ \lambda_k := \tilde{\lambda}_k^{(j)} \quad \text{for} \quad 0 \leqslant k \leqslant n-1$$

and introduce the norm $\|\cdot\|_C$ and scalar product $(\,,\,)_C$ by

$$\|x\|_C := \|C^{1/2}x\| , \quad (x,y)_C := x^T Cy .$$

Then by step 2b) of (III.1)

$$\text{(III.9)} \qquad y_o - y_n = \sum_{k=0}^{n-1} \lambda_k d_k = \sum_{k=0}^{n-1} \bar{\lambda}_k \bar{d}_k ,$$

where because of the exact line searches used

(III.10)
$$\bar{\lambda}_k := \bar{d}_k^T g_k = \bar{d}_k^T [g_o - \bar{\lambda}_o C\bar{d}_o - \dots - \bar{\lambda}_{k-1} C\bar{d}_{k-1}] = \lambda_k \sqrt{d_k^T C d_k}$$

so that because of $\|d_k\| = 1$ and (III.7)

(III.11)
$$|\lambda_k| \sqrt{m} = |\lambda_k| / \|C^{-1/2}\| \leq |\bar{\lambda}_k| \leq |\lambda_k| \, \|C^{1/2}\| = |\lambda_k| \cdot \sqrt{M}$$
$$\max_{0 \leq k \leq n-j-1} |\bar{\lambda}_k| \leq 2|\lambda_i| \sqrt{M} \leq 2|\bar{\lambda}_i| \cdot \sqrt{M/m} \; .$$

Because of the conjugacy properties of the $\bar{d}_k$ (see(III.3)), the vector $\bar{\lambda} := (\bar{\lambda}_o, \bar{\lambda}_1, \dots, \bar{\lambda}_{n-1})^T$ therefore satisfies the set of linear equations

(III.12)
$$L \cdot \bar{\lambda} = \bar{D}^T g_o$$

with the lower triangular matrix

(III.13)
$$L := \begin{bmatrix} l_{oo} & & & 0 \\ | & \ddots & & \\ | & & \ddots & \\ l_{n-1,0} & \cdots & & l_{n-1,n-1} \end{bmatrix} = \begin{bmatrix} L_{11} & 0 \\ L_{21} & \underbrace{I}_{j} \end{bmatrix} \}j$$

$$l_{\mu\nu} := \bar{d}_\mu^T C \bar{d}_\nu \; , \quad l_{kk} = 1 \; .$$

We partition $\bar{\lambda}$ and $\bar{D}^T g_o$ in the same way

$$\bar{\lambda} = \begin{pmatrix} \bar{\lambda}^{(1)} \\ \bar{\lambda}^{(2)} \end{pmatrix} \; , \quad \bar{D}^T g_o = \begin{pmatrix} \bar{D}_1^T g_o \\ \bar{D}_2^T g_o \end{pmatrix} = \begin{pmatrix} \bar{D}_1^T g_o \\ 0 \end{pmatrix}$$

where $\bar{\lambda}^{(2)}$, $\bar{D}_2^T g_o \in R^j$, and $\bar{D}_2^T g_o = 0$ vanishes since $y_o = \tilde{y}_o^{(j)}$ has been obtained by minimizing into the C-conjugate directions $\bar{d}_{n-j}, \bar{d}_{n-j+1}, \dots, \bar{d}_{n-1}$ in the previous line searches

$$\tilde{y}_{n-j+1}^{(j-1} \rightarrow \tilde{y}_{n-j+2}^{(j-1)} \rightarrow \dots \rightarrow \tilde{y}_n^{(j-1)} \rightarrow \tilde{y}_o^{(j)} = y_o \; .$$

It follows from (III.9), (III.12) that

$$y_o - y_n = \bar{D}\cdot\bar{\lambda} = \bar{D}L^{-1}\bar{D}^T g_o$$

and therefore

$$\|g_o\|/\|y_o - y_n\| \leq \|\bar{D}^{-1}\|^2\cdot\|L\| \quad .$$

Since the matrix $C^{1/2}\bar{D}$ has columns of euclidean length 1, it is easily seen from (III.13) that $\|L\| \leq n$, hence

$$\text{(III.14)} \qquad \|g_o\|/\|y_o - y_n\| \leq n\|\bar{D}^{-1}\|^2 \quad .$$

We next give an upper bound for $\tilde{D}_{j+1}$ in terms of $\tilde{D}_j$. If we again normalize the columns of $\tilde{D}_{j+1}$ with respect to $\| \ \|_C$ to length 1, we get a matrix $\bar{D}'$ with (see step 2d)α) of (III.1);

$$\bar{D}' := (\bar{d}_o,\ldots,\bar{d}_{i-1}, \bar{d}' , \bar{d}_{i+1},\ldots,\bar{d}_{n-1})\cdot P$$

where P is a suitable permutation and

$$\bar{d}' := (y_o - y_n)/\|y_o - y_n\|_C$$

$$= \alpha_o\bar{d}_o + \alpha_1\bar{d}_1 + \ldots + \alpha_i\bar{d}_i + \ldots + \alpha_{n-1}\bar{d}_{n-1}$$

$$\alpha_k := \bar{\lambda}_k/\|y_o - y_n\|_C \quad .$$

As $\alpha_i \neq 0$ because of the pivot selection (III.7), 2c") and $g_o \neq 0$, also $\bar{D}'$ is nonsingular and we have $\bar{D}' = \bar{D}\cdot G^{-1}\cdot P$, so that

$$\begin{aligned}\|(\bar{D}')^{-1}\| &\leq \|\bar{D}^{-1}\|\cdot\|P^{-1}\|\cdot\|G\| \\ \text{(III.15)} \qquad &= \|\bar{D}^{-1}\|\cdot\|G\| \quad ,\end{aligned}$$

where G is the Frobenius matrix

$$G = \begin{pmatrix} 1 & & & -\alpha_o/\alpha_i & & & \\ & \ddots & & \vdots & & & \\ & & 1 & \vdots & & & \\ & & & 1/\alpha_i & & & \\ & & & \vdots & 1 & & \\ & & & \vdots & & \ddots & \\ & & & -\alpha_{n-1}/\alpha_i & & & 1 \end{pmatrix} .$$

A bound for $\|G\|$ can be found as follows: Since $\alpha_k/\alpha_i = \bar{\lambda}_i$ and because of (III.11), we have

$$|\alpha_k/\alpha_i| \leq 2\sqrt{M/m} \quad k = 0,1,\ldots,n-j-1 . \tag{III.16}$$

By (III.12) and the particular structure of L (III.13) and $\bar{D}^T g_o$,

$$\bar{\lambda}^{(2)} = -L_{21}\bar{\lambda}^{(1)}$$

so that for $k \geq n-j$

$$\begin{aligned} |\bar{\lambda}_k| &\leq \|\bar{\lambda}^{(2)}\|_\infty \leq \|L_{21}\|_\infty \cdot \|\bar{\lambda}^{(1)}\|_\infty \\ &\leq 2|\bar{\lambda}_i| \cdot \|L_{21}\|_\infty \cdot \sqrt{M/m} \quad \text{(by (III.11))} \\ &\leq 2(n-j)\cdot|\bar{\lambda}_i| \cdot \sqrt{M/m} \end{aligned} \tag{III.17}$$

and therefore

$$|\alpha_k/\alpha_i| \leq 2(n-j)\sqrt{M/m} \quad \text{for } k=n-j,\ n-j+1,\ldots,n-1 . \tag{III.18}$$

Finally, by (III.9), (III.17), (III.11)

$$\begin{aligned} \|y_o-y_n\|_C &= \|C^{1/2}\bar{D}_1\bar{\lambda}^{(1)} + C^{1/2}\bar{D}_2\bar{\lambda}^{(2)}\|_2 \\ &\leq (n-j)\|\bar{\lambda}^{(1)}\|_\infty + j\|\bar{\lambda}^{(2)}\|_\infty \\ &\leq (n-j)\|\bar{\lambda}^{(1)}\|_\infty + j(n-j)\|\bar{\lambda}^{(1)}\|_\infty \\ &\leq 2(n-j)(j+1)\cdot|\bar{\lambda}_i|\cdot\sqrt{M/m} \end{aligned} \tag{III.19}$$

so that

$$|1/\alpha_i| = \|y_o-y_n\|_C/|\bar{\lambda}_i| \leq 2(n-j)(j+1)\sqrt{M/m} . \tag{III.20}$$

A tedious but elementary estimate of $\|G\|$ using (III.16)-(III.20) gives $\|G\| \leq 2n^2\sqrt{M/m}$ so that by (III.15) we have the recursive estimate

$$\|\bar{D}_{j+1}^{-1}\| \leq 2n^2\sqrt{M/m}\ \|\bar{D}_j^{-1}\|$$

where we have written $\bar{D}_{j+1}$ for $\bar{D}'$ and $\bar{D}_j$ for $\bar{D}$. Since $\bar{D}_j = \tilde{D}_j\Delta_j^{-1}$, where Δ_j is a diagonal matrix with

$$\|\Delta_j\| \leq \sqrt{M}, \quad \|\Delta_j^{-1}\| \leq 1/\sqrt{m}$$

and initially, $\tilde{D}_o = \Omega$ is orthogonal, we get

$$\|\tilde{D}_j^{-1}\| \leq (2n^2 \sqrt{M/m})^j \cdot \sqrt{M/m} .$$

Since $\tilde{D}_j$ has columns of euclidean length 1 ,

$$\|\tilde{D}_j\| \leq \sqrt{n} .$$

Together with(III.14) this proves the estimates a),b) of the theorem. As a byproduct of (III.19), we have

$$\|y_o - y_n\|_C \leq (n-j)(j+1) \cdot \|\bar{\lambda}^{(1)}\|_\infty$$

giving

$$\|\bar{\lambda}^{(1)}\|_\infty \geq \frac{\|y_o - y_n\|_C}{(n-j)(j+1)} \geq \|y_o - y_n\| \cdot \sqrt{m}/((n-j)(j+1)) \geq \frac{4\sqrt{m}}{(n+1)^2} \cdot \|y_o - y_n\|,$$

so that part c) of the theorem follows from part b) and (III.11). □

We apply the estimates of Theorem (III.8) to prove an analogue of Lemma (3.9) of [8,I]. The situation is more complicated than in [8,I], since the update function for D_{j+1} is discontinuous for certain arguments, as it depends on the outcome of the pivot selection in step 2c) of (III.1) and thereby on the function f to be minimized, and the whole history of (III.1). To describe the situation precisely and to circumvent these difficulties, let us now consider a fixed function $f(x)$ satisfying (A), (II.1), fixed starting values $x_o \in R^n$, Ω and fixed numbers $\alpha_j, \alpha_k^{(j)}$, α with $|\alpha_j| \leq \alpha$, $|\alpha_k^{(j)}| \leq \alpha$ specifying the asymptotic line searches (III.2) by setting

$$\mu_k^{(j)} := \alpha_k^{(j)} \|g_k^{(j)}\| , \quad \mu_j := \alpha_j \cdot \|g_j\| .$$

Then the indices $i_j := i(j)$ selected in (III.1),2c) and the iterates $x_j, y_k^{(j)}$, D_j of (III.1) when applied to f are well defined for $0 \leq j \leq j_o$, $j_o \leq n+1$ a certain integer. Next, denote by $\tilde{f}(x) = \frac{1}{2} x^T C x$ the fixed strictly convex quadratic function with $C := h(o)$ (see (II.1), (II.1')) associated with f and denote by $\tilde{x}_j$, $\tilde{y}_k^{(j)}$, $\tilde{D}_j$ the iterates obtained from the starting values $\tilde{x}_o := x_o$, $\tilde{D}_o = D_o = \Omega$ if the method (II.1) is modified as follows:

(III.21) a) <u>All line searches are exact.</u>
b) <u>Step 2c) of (III.1) is replaced by:</u>
2c") <u>For</u> $j < n$, <u>set</u> $i := i_j$.

That is, there is no pivot choice for $\tilde{f}$: it is dictated by the outcome of the pivot selection in step 2c) of (III.1) for the fixed function f. The following lemma provides recursive bounds for the differences $\Delta x_j := x_j - \tilde{x}_j$, $\Delta D_j := D_j - \tilde{D}_j$ in much the same way as the corresponding lemma (3.9) of [8,I]:

(III.22) **LEMMA:** <u>There are positive constants</u> $\xi_j, \delta_j, \epsilon_j$, $0 \leq j \leq j_o$, <u>which only depend on</u> Λ, m, M <u>and</u> α, <u>with the following properties: For all</u> $0 \leq j \leq j_o$ <u>and</u> $\|x_o\| \leq \epsilon_o$ <u>the inequalities</u>

(III.23)
a) $\|\Delta x_j\| \leq \xi_j \|x_o\|^2$
b) $\|\Delta D_j\| \leq \delta_j \|x_o\|^2 / \|x_\mu\|$, $\mu := \max(0, j-1)$

<u>imply the same inequalities with</u> j <u>replaced by</u> $j+1$, <u>provided</u>

(III.24) $x_j \neq 0$ <u>and</u> $\|x_o\|^2/\|x_j\| \leq \epsilon_{j+1}$.

The <u>proof</u> of Lemma (3.9) of [8,I] can be copied with only a few changes. The constants ξ_j, δ_j, ϵ_{j+1} are defined by induction on j. Set $\xi_o = \delta_o = 0$ and choose $\epsilon_o := \frac{\rho}{2}\sqrt{m/M}$. Then because of (II.6), $\|x_j\| \leq \rho$, $\|\tilde{x}\| \leq \rho$ for all j and the estimates (II.1) can be applied. Suppose that for some $j \geq 0$, ξ_ν, δ_ν, ϵ_ν have been defined for all $0 \leq \nu \leq j$, that (III.23) holds, and that for $\nu < j$ the step-sizes $\tilde{\lambda}_k^{(\nu)}$ determined during the application of (III.1), (III.21), to $\tilde{f}$ satisfy the estimates (compare (III.7))

(III.25) $$|\tilde{\lambda}_{i_\nu}^{(\nu)}| \geq \tfrac{1}{2} \max\{|\tilde{\lambda}_k^{(\nu)}| : 0 \leq k \leq n-\nu-1\}.$$

We have to define ϵ_{j+1}, ξ_{j+1}, δ_{j+1} with the properties of the lemma. As j is fixed for the moment, we may use the simplified notation

$$y_k := y_k^{(j)} \; , \; \tilde{y}_k := \tilde{y}_k^{(j)} \; , \; \lambda_k = \lambda_k^{(j)} \; , \; \tilde{\lambda}_k = \tilde{\lambda}_k^{(j)} \; .$$

A first requirement for ϵ_{j+1} is

$$\epsilon_{j+1}\xi_j \leq \tfrac{1}{2} \; . \tag{III.26}$$

Then as in the proof of (3.9) of [8,I]

$$\tfrac{1}{2}\|x_j\| \leq \|\tilde{x}_j\| \leq 2\|x_j\| \; . \tag{III.27}$$

Also, by definition

$$\|\Delta y_o\| = \|y_o - \tilde{y}_o\| \leq \eta_o \cdot \|x_o\|^2 \; , \quad \eta_o := \xi_j$$

$$\tfrac{1}{2}\|y_o\| \leq \|\tilde{y}_o\| \leq 2\|y_o\|$$

The differences $\Delta y_k := y_k - \tilde{y}_k$, $k=0,1,\ldots,n$ can now be estimated recursively using proposition (II.8), (II.6) and the induction hypothesis:

$$\|\Delta y_{k+1}\| \leq \sigma_1\|\Delta y_k\| + \sigma_2\|\tilde{y}_k\| \; \|\Delta D_j\| + \sigma_3\|\Delta g_k\| + \sigma_4\|y_k\|^2 + \sigma_5\|y_k\|\,\|\tilde{y}_k\|$$

$$\leq \eta_{k+1}\|x_o\|^2$$

for some constants η_{k+1}, δ_i, β_g of "type β", i.e..., constants only depending on Λ, m, M and α. Here we have used that

$$\|\Delta y_k\| \leq \eta_k\|x_o\|^2$$

$$\|\tilde{y}_k\| \leq 2\sqrt{\tfrac{M}{m}}\|\tilde{y}_o\| = 2\sqrt{\tfrac{M}{m}}\,\|\tilde{x}_j\| \leq 4\sqrt{\tfrac{M}{m}}\,\|x_j\| = O(\|x_o\|)$$

$$\|y_k\| \leq \|\tilde{y}_o\| + \|\Delta y_k\| = O(\|x_o\|)$$

$$\|\tilde{y}_k\| \; \|\Delta D_j\| = O(\|x_o\|^2)$$

$$\|\Delta g_k\| = \|g(y_k) - \tilde{g}(\tilde{y}_k)\| \leq \beta_g\|x_o\|^2$$

$$\leq \|g(y_k) - \tilde{g}(y_k)\| + \|\tilde{g}(y_k) - \tilde{g}(\tilde{y}_k)\| \leq \tfrac{\Lambda}{2}\|y_k\|^2 + 2M\|\Delta y_k\| = O(\|x_o\|^2) \; .$$

Hence, the vectors $p := y_o - y_n$, $\tilde{p} := \tilde{y}_o - \tilde{y}_n$ satisfy a bound

$$\|\Delta p\| = \|p - \tilde{p}\| \leq \beta_p \|x_o\|^2 \tag{III.28}$$

for another constant $\beta_p = \eta_o + \eta_n$ of type β. Also it can be shown that the differences $\Delta\lambda_k = \lambda_k - \tilde{\lambda}_k$ are bounded

$$|\Delta\lambda_k| \leq \beta_\lambda \|x_o\|^2 \quad , \; k=0,1,\ldots,n-1 \tag{III.29}$$

with some β_λ of type β. Therefore, by definition of the index $i = i_j$ selected in step 2c) of (III.1) for f

$$|\tilde{\lambda}_k| - |\Delta\lambda_k| \leq |\lambda_k| \leq |\lambda_i| \leq |\tilde{\lambda}_i| + |\Delta\lambda_i|, \; k=0,1,\ldots,n-j-1,$$

so that by (III.29)

$$\max_{0\leq k\leq n-j-1} |\tilde{\lambda}_k| - 2\beta_\lambda \|x_o\|^2 \leq |\tilde{\lambda}_i| \; . \tag{III.30}$$

However, by induction hypothesis (III.25), theorem (III.8),c) can be applied. This gives the estimate

$$\max_{0\leq k\leq n-j-1} |\tilde{\lambda}_k| \geq \|C\tilde{x}_j\|/\beta \geq m\|x_j\|/(2\beta) = \|x_j\|/\bar{\beta}_\lambda$$

for some constants $\bar{\beta}_\lambda$ of type β. Hence, by (III.30), if

$$2\beta_\lambda \|x_o\|^2 \leq \|x_j\|/(2\bar{\beta}_\lambda) \quad ,$$

or if

$$4\varepsilon_{j+1} \cdot \beta_\lambda \bar{\beta}_\lambda < 1 \tag{III.31}$$

then

$$\max_{0\leq k\leq n-j-1} |\tilde{\lambda}_k| \leq 2|\tilde{\lambda}_i| \tag{III.32}$$

and theorem (III.8)b) can be applied also for the index j under consideration. This gives the bounds

$$\frac{m}{2}\|x_j\| \leq m\|\tilde{x}_j\| \leq \|C\tilde{x}_j\| \leq \beta\|\tilde{p}\|$$

and therefore

$$\|x_j\|/\|\tilde{p}\| \leq \beta_1 \tag{III.33}$$

for some constant β_1 of type β. Now, the difference ΔD_{j+1} can be estimated. By construction of D_{j+1} and (III.21) we have

$$\|\Delta D_{j+1}\| \leq \|\Delta D_j\| + \|\Delta d\| \leq \delta_j \cdot \sqrt{M/m}\ \|x_o\|^2/\|x_j\| + \|\Delta d\|$$

where $d := p/\|p\|$, $\tilde{d} := \tilde{p}/\|\tilde{p}\|$. If

$$\beta_1 \beta_p \epsilon_{j+1} \leq 1/2 \tag{III.34}$$

then by (III.28), (III.33)

$$\|p\| \geq \|\tilde{p}\| \left(1 - \frac{\|\Delta p\|}{\|\tilde{p}\|}\right) \geq \|\tilde{p}\| (1 - \beta_1 \beta_p \epsilon_{j+1}) \geq \tfrac{1}{2} \|\tilde{p}\| .$$

Therefore

$$\|\Delta d\| \leq \frac{\|\Delta p\|}{\|p\|} + \|\tilde{p}\| \left(\frac{1}{\|p\|} - \frac{1}{\|\tilde{p}\|}\right) \leq \frac{2\|\Delta p\|}{\|p\|} \leq 4 \frac{\|\Delta p\|}{\|\tilde{p}\|}$$

$$\leq 4\beta_p \beta_1 \|x_o\|^2/\|x_j\|$$

so that

$$\|\Delta D_{j+1}\| \leq \delta_{j+1} \|x_o\|^2/\|x_j\|$$

for the constants $\delta_{j+1} := \delta_j \sqrt{M/m} + 4\beta_p \beta_1$ of type β.

Now a bound for Δx_{j+1} is easily obtained from Proposition (II.8): By definition of x_{j+1}, $\tilde{x}_{j+1}$ (step 2dβ) of (III.11)), we have by (II.8)

$$\|\Delta x_{j+1}\| \leq \sigma_1 \|\Delta x_j\| + \sigma_2 \|\tilde{x}_j\| \cdot \|\Delta D_{j+1}\| + \sigma_3 \|\Delta g_j\| + \sigma 4 \|x_j\|^2 + \sigma_5 \|x_j\| \cdot \|\tilde{x}_j\|,$$

$$\Delta g_j := g(x_j) - \tilde{g}(\tilde{x}_j) ,$$

for some constants σ_k of type β. By what we have proved so far, this implies a bound of the desired form

$$\|\Delta x_{j+1}\| \leq \xi_{j+1} \|x_o\|^2$$

for some ξ_{j+1} of type β . So, the constants ξ_{j+1} , δ_{j+1} and (compare (III.26), (III.31), (III.34))

$$\epsilon_{j+1} := \min\left\{\frac{1}{2\xi_j}, \frac{1}{4\beta_\lambda\bar{\beta}_\lambda}, \frac{1}{2\beta_1\beta_p}\right\}$$

have the required properties. This completes the proof of lemma (III.22). □

The following theorem follows in much the same way from lemma (III.22) as in [8,I] theorem (III.8) follows from lemma (III.9). Therefore, its proof will be omitted:

(III.35) **THEOREM:** Suppose f satisfies (A), (II.1), and suppose that the method (III.1) is applied to f using asymptotically perfect line searches (III.2). Then there are positive constants ϵ , c only depending on m, M, Λ and α such that for all starting values $x_o \in R^n$ with $\|x_o\| \leq \epsilon$ and all orthogonal starting matrices Ω , the method (III.1) generates iterates x_j, $j=0,1,\ldots,j_o \leq n+1$ with the following property: There is an index $\mu \leq \min\{n,j_o\}$ such that

$$\|x_\mu\| \leq c\cdot\|x_o\|^2 .$$

The method (III.1) when used with restarts therefore is at least n-step quadratically convergent, if only the transitions $x_j \to x_{j+1}$ between the main iterates are counted as steps.

IV. PARALLEL PROJECTION METHODS

Powell's method (III.1) is essentially a sequential method, but there are related minimization methods which are suitable for parallel computers. Methods of this type have been considered by several authors, among others by Chazan and Miranker [2] and by Sloboda [6]. We again describe only a basic cycle of these methods, leading from an initial approximation x_o for the minimum of a function $f\colon R^n \to R$ to better approximations x_j, $j=0,1,\ldots,n$. Of course, using restart techniques x_n can be further improved by taking it as a starting value for a new cycle of the method. We describe the method in a form considered by Sloboda [6].

(IV.1) <u>Given</u> $x_o \in R^n$ <u>and</u> n <u>linearly independent vectors</u> $d_1, d_2, \ldots, d_n \in R^n$, <u>which are assumed to be normalized</u>, $\|d_i\| = 1$.

1. <u>If</u> $\|g(x_o)\| = 0$ <u>stop</u>: x_o <u>is a stationary point of</u> f. <u>Otherwise choose a suitable number</u> $\sigma > 0$ <u>and set</u>

$$y_o^{(o)} := x_o \;,$$

$$\left.\begin{aligned} v_o^{(k)} &:= \sigma \cdot d_k \;, \\ y_o^{(k)} &:= y_o^{(o)} + v_o^{(k)} \end{aligned}\right\} \quad \underline{\text{for}}\ \ k=1,2,\ldots,n \;.$$

2. <u>For</u> $j := 0,1,\ldots,n-1$ <u>do</u>

   a) <u>set</u> $v_j := v_j^{(j+1)} := y_j^{(j+1)} - y_j^{(j)} = y_j^{(j+1)} - x_j$.

   b) <u>For</u> $k := j+1, j+2, \ldots, n$ <u>compute</u> $\lambda = \lambda_j^{(k)}$ <u>to minimize</u> $f(y_j^{(k)} - \lambda v_j)$ <u>approximately and set</u>

$$y_{j+1}^{(k)} := y_j^{(k)} - \lambda_j^{(k)} v_j \quad .$$

   c) <u>Set</u>

$$x_{j+1} := y_{j+1}^{(j+1)}$$

$$v_{j+1}^{(k)} := y_{j+1}^{(k)} - y_{j+1}^{(j+1)} \quad \underline{\text{for}}\ \ k=j+2,\ldots,n \;.$$

The line searches in step 2b) are again called <u>asymptotically perfect</u> if there is a constant $\alpha \geq 0$ such that

$$\text{(IV.2)} \quad g_{j+1}^{(k)\,T} v_j = \mu_j^{(k)} \cdot g_j^{(k)\,T} v_j \;, \quad |\mu_j^{(k)}| \leq \alpha \, \|g_k^{(j)}\| \;,$$

where $g_j^{(k)} := g(y_j^{(k)})$. In addition, we require

$$\text{(IV.3)} \quad f(y_{j+1}^{(k)}) \leq f(y_j^{(k)}) \ , \ k=j+1,\ldots,n$$

for the line searches in step 2b) of (IV.1) .

By V_j, $j=0,1,\ldots,n-1$, we denote the $n\times n$-matrices

$$\text{(IV.4)} \quad V_j := (v_o^{(1)}, v_1^{(2)},\ldots,v_j^{(j+1)}, v_j^{(j+2)},\ldots,v_j^{(n)})$$
$$= (v_o, v_1,\ldots,v_j, v_j^{(j+2)},\ldots, v_j^{(n)}), \ j=0,1,\ldots,n-1$$
$$V_o = \sigma\cdot D \ , \ D:= (d_1,d_2,\ldots,d_n)$$

associated with the vectors $v_j^{(k)}$ generated by (IV.1) . In step 1) of (IV.1) the constant $\sigma > 0$ is chosen such that

$$\text{(IV.5)} \quad \min\{h,\sigma_1\|g(x_o)\|\} \leq \sigma \leq \min\{h, \sigma_2\|g(x_o)\|\} ,$$

where $h > 0$, $0 < \sigma_1 < \sigma_2$ are fixed numbers. This is possible, if x_o is not stationary.

In view of the basic assumption (II.1), any such choice ensures that the auxiliary initial points $y_o^{(k)}$, $k=1,2,\ldots,n$, in (IV.1) are close to the local minimum $\bar{x} = 0$ of f , if $x_o = y_o^{(o)}$ is close to $\bar{x}$:

There is a constant β only depending on f and α such that

$$\|y_o^{(k)}\| \leq \beta \ \|x_o\| \ , \ k=1,2,\ldots,n .$$

Note further that the matrices V_j (IV.4), $j=0,1,\ldots,n-1$ are nonsingular. This follows easily from the recursion formula

$$v_{j+1}^{(k)} = y_{j+1}^{(k)} - y_{j+1}^{(j+1)} = y_j^{(k)} - \lambda_j^{(k)} v_j - (y_j^{(j+1)} - \lambda_j^{(j+1)} v_j) - y_j^{(j)} + y_j^{(j)}$$

$$\text{(IV.6)} \quad = v_j^{(k)} + (\lambda_j^{(j+1)} - \lambda_j^{(k)} - 1)v_j \ , \ k=j+2, j+3,\ldots,n ,$$

and the nonsingularity of V_o .

The behaviour of (IV.1) for quadratic functions $\tilde{f}$ is well known, and the main result is contained in the following theorem, which is stated without proof (see e.g. Chazan, Miranker [2], Sloboda [6]) Hestenes [3]:

(IV.7) **THEOREM:** Let $\tilde{f}(x) = \frac{1}{2} x^T C x$ be a strictly convex quadratic function satisfying (II.1) and suppose that (IV.1) as applied to $\tilde{f}$ using only exact line searches in step 2b) of (IV.1). Then for arbitrary starting values $\tilde{x}_o \in R^n$, $\sigma > 0$ and $D = (d_1, \ldots, d_n)$ a nonsingular matrix, (IV.1) generates vectors $\tilde{x}_j$, $\tilde{y}_j^{(k)}$, $\tilde{v}_j^{(k)}$ and matrices V_j such that

1) $\tilde{x}_n = 0$ is the minimum of $\tilde{f}$

2) The vectors $\tilde{v}_j = \tilde{v}_j^{(j+1)}$, $j=0,1,\ldots,n-1$ are C-conjugate

$$\tilde{v}_j^T C \tilde{v}_k = 0 \quad \text{for} \quad j \neq k \,, \quad \tilde{v}_j^T C \tilde{v}_j > 0 \,.$$

We use this result, in order to prove some estimates for the directions $\tilde{v}_j^{(k)}$ to be used later on

(IV.8) **LEMMA:** Under the hypotheses of theorem (IV.7), the vectors $\tilde{v}_j^{(k)}$ can be expressed in terms of the initial vectors $\tilde{v}_o^{(k)}$ and $\tilde{v}_s = \tilde{v}_s^{(s+1)}$, $s \leq j-1$, by

$$\text{(IV.9)} \quad \tilde{v}_j^{(k)} = \tilde{v}_o^{(k)} - \sum_{s=0}^{j-1} \frac{\tilde{v}_s^T C \tilde{v}_o^{(k)}}{\tilde{v}_s^T C \tilde{v}_s} \cdot \tilde{v}_s \,, \quad j=0,1,\ldots,n-1, \quad k=j+1,j+2,\ldots,n \,.$$

The vectors $\tilde{v}_j^{(k)}$ are linear functions of $\sigma > 0$. there are positive constants β, β_r only depending on C (via m and M, see (II.1)) and D such that for all $\tilde{x}_o \in R^n$, $0 < \sigma \leq \beta$ the following estimates hold:

$$\text{(IV.10)} \quad \begin{aligned} &\beta_1 \cdot \sigma \leq \|\tilde{v}_j\| \leq \beta_2 \, \sigma \\ &\|\tilde{V}_j\| \leq \beta_3 \, \sigma \,, \quad \|\tilde{V}_j^{-1}\| \leq \beta_4 / \sigma \end{aligned}$$

<u>for</u> $j = 0,1,\ldots,n-1$.

<u>Proof:</u> Since the line search in step 2b) of (IV.1) is exact, we have that

$$\tilde{v}_j^T\, C(\tilde{y}_j^{(k)} - \tilde{\lambda}_j^{(k)}\tilde{v}_j) = 0$$

so that

$$\tilde{\lambda}_j^{(k)} = \frac{\tilde{v}_j^T\, C\tilde{y}_j^{(k)}}{\tilde{v}_j^T\, C\tilde{v}_j} = [\tilde{v}_j^T\, C\tilde{y}_j^{(j)} + \tilde{v}_j^T\, C\tilde{v}_j^{(k)}]/\tilde{v}_j^T\, C\tilde{v}_j \quad .$$

Inserting this expression for $\tilde{\lambda}_j^{(k)}$ into (IV.6) gives the recursion

$$\tilde{v}_{j+1}^{(k)} = \tilde{v}_j^{(k)} - \frac{\tilde{v}_j^T\, C\tilde{v}_j^{(k)}}{\tilde{v}_j^T\, C\tilde{v}_j}\, \tilde{v}_j \ , \ j=0,1,\ldots,n-1 \ ,$$

which proves (IV.9). In order to prove (IV.10) we first introduce the abbreviations

$$r_{j+1,j+1} := \sqrt{\tilde{v}_j^T\, C\tilde{v}_j} \qquad \bar{v}_j := \frac{1}{r_{j+1,j+1}} \cdot \tilde{v}_j$$

$$r_{s+1,k} := \frac{\tilde{v}_s^T\, C\tilde{v}_o^{(k)}}{\sqrt{\tilde{v}_s^T\, C\tilde{v}_s}} = \bar{v}_s^T\, C\tilde{v}_o^{(k)} \quad ,$$

$$R := \begin{pmatrix} r_{11} & \cdots & r_{1n} \\ & \ddots & \vdots \\ 0 & & r_{nn} \end{pmatrix} \quad , \quad R_j := \begin{pmatrix} r_{11} & r_{1j} & & r_{1n} \\ & r_{jj} & & r_{jn} \\ & & 1 & 0 \\ & & & \vdots \\ 0 & & & 1 \end{pmatrix}$$

$$\bar{V}_{j-1} := (\bar{v}_o, \bar{v}_1, \ldots, \bar{v}_{j-1}, \tilde{v}_{j-1}^{(j+1)} \tilde{v}_{j-1}^{(j+2)}, \ldots, \tilde{v}_{j-1}^{(n)}) \quad ,$$

so that (IV.9) is equivalent to

$$\tilde{v}_o^{(k)} = \tilde{v}_j^{(k)} + \sum_{s=1}^{j} r_{s,k} \cdot \bar{v}_{s-1}, \quad k=j+1,j+2,\ldots,n, \quad j=0,1,\ldots,n-1$$

or to the matric equation

$$\text{(IV.11)} \quad \sigma D = \tilde{V}_o = (\tilde{v}_o^{(1)}, \ldots, \tilde{v}_o^{(n)}) = \bar{V}_{j-1} \cdot R_j \ , \ j=1,2,\ldots,n-1,$$

in particular for $j = n$

$$\sigma D = \tilde{V}_o = \bar{V}_{n-1} \cdot R \quad .$$

Now, because of the C-conjugancy of the $\tilde{v}_j$

$$\bar{V}_{n-1}^T C \, \bar{V}_{n-1} = I$$

the matrix $Q := C^{\frac{1}{2}} \bar{V}_{n-1}$ is an orthogonal matrix so that

$$\|\bar{V}_{n-1}\| \leq \|C^{-\frac{1}{2}}\| \leq 1/\sqrt{m}$$

$$\|V_{n-1}^{-1}\| \leq \|C^{\frac{1}{2}}\| \leq \sqrt{M}$$

(IV.12)
$$\|R\| = \sigma \|\bar{V}_{n-1}^{-1} \, D\| \leq \sigma \sqrt{M} \, \|D\|$$

$$\|R^{-1}\| = \frac{1}{\sigma} \|D^{-1} \, \bar{V}_{n-1}\| \leq \|D^{-1}\|/(\sigma \sqrt{m}) .$$

In particular it follows from these estimates

(IV.13)
$$\tilde{v}_j^T \, C \tilde{v}_j = r_{j+1,j+1}^2 \leq \|R\|^2 \leq \sigma^2 \, M \|D\|^2$$

$$r_{j+1,j+1}^2 \geq 1/\|R^{-1}\|^2 \geq \sigma^2 m/\|D^{-1}\|^2 \quad ,$$

and for the $n \times n$-matrix $\Delta_j := \text{diag}(r_{11}, r_{22}, \ldots, r_{j,j}, 1, \ldots, 1)$

(IV.14)
$$\|\Delta_j\|^2 \leq \max(\sigma^2 M \, \|D\|^2 \, , \, 1)$$

$$\|\Delta_j^{-1}\|^2 \leq \max\left(\frac{1}{\sigma^2 m} \, \|D^{-1}\|^2 \, , \, 1\right) \quad .$$

Hence, by (IV.10)

$$\tilde{V}_{j-1} = \bar{V}_{j-1} \cdot \Delta_j = \sigma \, D R_j^{-1} \Delta_j$$

(IV.15)
$$\|\tilde{V}_{j-1}\| \leq \sigma \, \|D\| \cdot \|R_j^{-1}\| \, \|\Delta_j\|$$

$$\|\tilde{V}_{j-1}^{-1}\| \leq \frac{1}{\sigma} \, \|D^{-1}\| \cdot \|R_j\| \, \| \, \Delta_j^{-1}\|$$

The estimates (IV.10) now follow directly from (IV.12) - (IV.15) and the easily established bounds

$$\|R_j\| \leq \|R\| + 1$$

$$\|R_j^{-1}\| \leq \|R^{-1}\| \cdot \|R\| + 1 \, ,$$

$$\frac{1}{M}\, r_{j+1,j+1}^2 \leqslant \|\tilde{v}_j\|^2 \leqslant \frac{1}{m}\, r_{j+1,j+1}^2 \quad . \quad \square$$

Now, we can prove our main theorem on the rate of convergence of algorithm (IV.1):

(IV.16) **THEOREM:** <u>Suppose that the function</u> f <u>satisfies</u> (II.1) <u>and</u> (IV.1) <u>is applied to minimize</u> f <u>using asymptotically perfect line searches satisfying</u> (IV.2), (IV.III) <u>with a certain constant</u> $\alpha \geqslant 0$ <u>and</u> σ <u>satisfying</u> (IV.5). <u>Then for every nonsingular</u> $n \times n$-<u>matrix</u> $D = (d_1, d_2, \ldots, d_n)$ <u>with normalized columns</u> $\|d_i\| = 1$ <u>there are constants</u> $\varepsilon > 0$, $\gamma \geqslant 0$ <u>only depending on</u> f (<u>via</u> Λ, m, M) , α <u>and</u> σ_1, σ_2 <u>such that for all starting points</u> $\|x_0\| \leqslant \varepsilon$, <u>all</u> σ <u>with</u> (IV.5) <u>the n-th iterate</u> x_n <u>generated by</u> (IV.1) <u>satisfy the estimate</u>

$$\|x_n\| \leqslant \gamma\, \|x_0\|^2 \quad .$$

As a consequence, the algorithm (IV.1) used with n-periodic restarts is n-step square convergent.

<u>Proof:</u> Let $\alpha \geqslant 0$, $h > 0$, $0 < \sigma_1 < \sigma_2$ and let $D = (d_1, \ldots, d_n)$ be an arbitrary fixed nonsingular matrix. Denote by x_j , v_j, $v_j^{(k)}$, $y_j^{(k)}$ the vectors generated by (IV.1) from any starting point x_0 , if σ satisfies (IV.5) and (IV.1) is applied to minimize f using asymptotically perfect line searches satisfying (IV.2), (IV.3). Denote by $\tilde{x}_j$, $\tilde{v}_j$, $\tilde{v}_j^{(k)}$, $\tilde{y}_j^{(k)}$ the corresponding quantities generated by (IV.1) from the same starting values $\tilde{D} = D$, $\tilde{x}_0 = x_0$, $\tilde{\sigma} = \sigma$, if (IV.1) is applied to the minimization of the quadratic function $\tilde{f}(x) = \frac{1}{2} x^T C x$ (see(II.1)) approximating f up to second order terms and only exact line searches are used in (IV.1). Then $\tilde{x}_n = 0$ by (IV.7). We shall estimate the differences $\Delta x_j := x_j - \tilde{x}_j, \ldots,$ $\Delta y_j^{(k)} := y_j^{(k)} - \tilde{y}_j^{(k)}$ recursively and we shall show that for sufficiently small $\varepsilon > 0$ and $\|x_0\| \leqslant \varepsilon, \Delta x_n = x_n - \tilde{x}_n = x_n$ satisfies a bound $\|\Delta x_n\| = \|x_n\| \leqslant \gamma\, \|x_0\|^2$ of the type

asserted in the theorem. In particular, we prove by induction on j , that there are constants $\beta^{(j)}$ only depending on f (via Λ, m, M) and σ_1, σ_2 such that for $j \geq 0$, $k=j,j+1,\ldots,n$

$$\text{(IV.17)} \quad \begin{aligned} &\text{a)} \quad \|\Delta y_j^{(k)}\| \leq \beta^{(j)} \|x_o\|^2 \\ &\text{b)} \quad \|\Delta v_j^{(k)}\| \leq 2\beta^{(j)} \|x_o\|^2 \end{aligned}$$

so that in particular for $k = j$

$$\|\Delta x_j\| = \|\Delta y_j^{(j)}\| \leq \beta^{(j)} \|x_o\|^2 \quad ,$$

provided $\|x_o\| \leq \varepsilon$ is small enough. For the proof set initially $\beta^{(o)} = 0$ and let $0 < \varepsilon_o \leq \rho$ be sufficiently small, so that the assertions of (II.1) and (II.6) hold. A first requirement for $\varepsilon > 0$ then is

$$\text{(IV.18)} \qquad \varepsilon \leq \tfrac{1}{2} \sqrt{\frac{m}{M}} \; \frac{\varepsilon_o}{1+2\sigma_2 M}$$

Then by (IV.5) and (II.1) for $\|x_o\| \leq \varepsilon$

$$0 < \sigma \leq \sigma_2 \; \|g(x_o)\| \leq \sigma_2 \cdot 2M \|x_o\|$$

$$\|y_o^{(k)}\| \leq \|x_o\| + \|v_o^{(k)}\| = \|x_o\| + \sigma \leq (1+2\sigma_2 M)\|x_o\|, \; k=1,2,\ldots,n \; ,$$

so that for $j \geq 0$, $k=j+1,\ldots,n$ because of (IV.3),(II.6)

$$\text{(IV.19)} \quad \begin{aligned} \|y_j^{(k)}\| &\leq 2\sqrt{\frac{M}{m}} \|y_o^{(k)}\| \leq 2\sqrt{\frac{M}{m}} \; (1 + 2\sigma_2 M) \|x_o\| \\ &= \beta \; \|x_o\| \leq \varepsilon_o \\ \|x_j\| &= \|y_j^{(j+1)}\| \leq \beta \; \|x_o\| \leq \varepsilon_o \end{aligned}$$

for some constant β of the type required.

Suppose inductively, that (IV.17) holds up to a certain $j \geq 0$. Then by(**IV.**17)b), (IV.5) and the estimate (IV.10) of Lemma (IV.8),

$$\|\Delta v_j\| / \|\tilde{v}_j\| \leq \frac{2\beta^{(j)} \; \|x_o\|^2}{\beta_1 \sigma_1 \|g(x_o)\|} \leq \tfrac{1}{2} \; \beta_s^{(j)} \|x_o\| \leq \tfrac{1}{2} \; \beta_s^{(j)} \cdot \varepsilon$$

for some constant $\beta_s^{(j)}$ independent of x_o. Hence, if in addition to (IV.18),

$$\epsilon \leqslant 1/\beta_s^{(j)} \tag{IV.20}$$

then

$$\|\Delta v_j\|/\|\tilde{v}_j\| \leqslant \tfrac{1}{2}$$

and therefore

$$\|v_j\| \geqslant \|\tilde{v}_j\| - \|\Delta v_j\| \geqslant \tfrac{1}{2}\,\|\tilde{v}_j\| > 0 \,.$$

Hence, the normalized search directions

$$s_j = \frac{v_j}{\|v_j\|}\,, \quad \tilde{s}_j := \frac{\tilde{v}_j}{\|\tilde{v}_j\|}$$

are well defined and we have the estimate

$$\|\Delta s_j\| = \|s_j - \tilde{s}_j\| = \left\| \frac{\Delta v_j \cdot \|v_j\| - v_j(\|v_j\| - \|\tilde{v}_j\|)}{\|v_j\| \cdot \|\tilde{v}_j\|} \right\|$$

$$\leqslant 2 \cdot \frac{\|\Delta v_j\|}{\|\tilde{v}_j\|} \leqslant \beta_s^{(j)}\,\|x_o\| \tag{IV.21}$$

Because of (IV.18), (IV.19), (IV.2), (IV.3), all hypotheses of (II.8) are satisfied and using the definitions of $y_{j+1}^{(k)}, \tilde{y}_{j+1}^{(k)}$, $k=j+1,\ldots,n$, (II.8) gives the bound

$$\begin{aligned} \|\Delta y_{j+1}^{(k)}\| \leqslant{}& \sigma_1\,\|\Delta y_j^{(k)}\| + \sigma_2\|\tilde{y}_j^{(k)}\|\cdot\|\Delta s_j\| + \sigma_3\|\Delta g_j^{(k)}\| + \\ &+ \sigma_4\|y_j^{(k)}\|^2 + \sigma_5\|y_j^{(k)}\|\,\|\tilde{y}_j^{(k)}\| \\ \Delta g_j^{(k)} :={}& g(y_j^{(k)}) - \tilde{g}(\tilde{y}_j^{(k)})\quad, \end{aligned} \tag{IV.22}$$

for some constants σ_i only depending on m, M, Λ and α.

Since by (II.1), (IV.17), (IV.19)

$$\|\Delta g_j^{(k)}\| \leq \|g(y_j^{(k)}) - \tilde{g}(y_j^{(k)})\| + \|\tilde{g}(y_j^{(k)}) - \tilde{g}(\tilde{y}_j^{(k)})\|$$

$$\leq \frac{\Lambda}{2} \|y_j^{(k)}\|^2 + M \cdot \|\Delta y_j^{(k)}\|$$

$$\leq (\frac{\Lambda}{2} \beta^2 + M\beta^{(j)}) \|x_o\|^2 =: \beta_g^{(j)} \|x_o\|^2 ,$$

the induction hypothesis (IV.17), (IV.19) and (IV.21) applied to (IV.22) give

$$\|\Delta y_{j+1}^{(k)}\| \leq [\delta_1 \beta^{(j)} + \delta_2 \cdot \beta\, \beta_s^{(j)} + \delta_3 \beta_g^{(j)} + (\delta_4 + \delta_5) \beta^2] \|x_o\|^2$$

$$=: \beta^{(j+1)} \|x_o\|^2 . \tag{IV.23}$$

This establishes (IV.17)a) for j+1. Moreover, by the definition of

$$v_{j+1}^{(k)} = y_{j+1}^{(k)} - y_{j+1}^{(j+1)} ,$$

(IV.23) implies immediately

$$\|\Delta v_{j+1}^{(k)}\| \leq \|\Delta y_{j+1}^{(k)}\| + \|\Delta y_{j+1}^{(j+1)}\|$$

$$\leq 2\beta^{(j+1)} \|x_o\|^2 .$$

This completes the inductive definition of the constants $\beta^{(j)}$. Clearly, the constants

$$\gamma := \beta^{(n)}$$

$$\epsilon := \min\{\tfrac{1}{2} \sqrt{\tfrac{m}{M}} \frac{\epsilon_o}{1 + \sigma_2 M} , \frac{1}{\beta_s^{(j)}} \,/\, j = 1,2,\ldots,n-1\}$$

have the properties asserted in the theorem. □

REFERENCES

1. R. Brent, Algorithms for minimization without derivatives, Englewood Cliffs, N.J.: Prentice Hall, 1973.

2. D. Chazan and W.L. Miranker, A nongradient and parallel algorithm for unconstrained minimization, SIAM J. Control 8, 1970, 207-214.

3. M. Hestenes, Conjugate direction methods in optimization, New York: Springer, 1980.

4. L. Kaufmann (1975), Unpublished manuscript, private communication.

5. M.J.D. Powell, An efficient method for finding the minimum of a function of several variables without calculating derivatives, Comput.J.7, 1964, 155-162.

6. F. Sloboda, Nonlinear iterative methods and parallel computation, Applikace Mat. 21, 1976, 252-262.

7. ________, A conjugate directions method and its application. p. 147-157 in : J. Stoer (ed.), Proceedings of the 8th IFIP-Conference on Optimization Techniques, Würzburg, 1977, Lecture Notes in Control and Information Sciences Vol.7 New York. Springer, 1978.

8. J. Stoer, On the relation between quadratic termination and convergence properties of minimization algorithms, Part I: Theory, Numer. Math. 28, 1977, 343-366. Part II (together with P. Baptist): Applications, Numer. Math. 28, 1977, 367-391.

9. C. Sutti, A new method for nonconstrained minimization without derivatives p.277-289 in: Towards global optimization (L.C.W. Dixon and G.P. Szegö eds.), North Holland, 1975.

10. ________, Convergence proof of minimization algorithms for nonconvex functions, JOTA 23, 1977, 203-210.

11. P.L. Toint, F.M. Callier, On the accelerating properties of an algorithm for function minimization without calculating derivatives, JOTA 23, 1977, 531-547.

12. P.L. Toint, F.M. Callier, On the uniform nonsingularity of matrices of search directions and the rates of convergence in minimization algorithms, JOTA 23, 1977, 511-529.

13. W.I. Zangwill, Minimizing a function with calculating derivatives, Comput. J. 10, 293-296.

III. APPLICATIONS

OPTIMAL RECONSTRUCTION OF SURFACES USING PARAMETRIC SPLINE FUNCTIONS

Philippe Cinquin

Department d'Informatique Medicale
Center Hospitalier Regional
Dijon, France

I. INTRODUCTION

The aim of this paper is to extend the well known properties of interpolation and smoothing spline functions to the case of multi-dimensional parametric interpolating and smoothing.

This generalization proved necessary to solve the following problem set to us by orthopaedist surgeons :

P: using a few X-ray photographies, define the shape of the medullary canal of the femur.

This problem arises in the automatic design of individually adapted hip prosthesis. The radiological analysis of this problem enables us to state it in the following way :

P_1: define a surface S from the knowledge of some elements tangent to that surface.

There comes the difficult point: indeed the usual interpolating and smoothing surfaces are defined by a set of values of the unknown function at the mesh points of a given network ; the lack of such a network leads us to use a parametric representation for the surface.

Our problem then reads :

P_2 Estimate a function $x : \mathbb{R}^2 \supset T \longrightarrow \mathbb{R}^3,\ (t_1,t_2) \longrightarrow x(t_1,t_2)$ with minimization properties to be precised and satisfying :

$$L_i(x) = z_i \qquad (i = 1,\dots,n) \quad \text{where } z_i (i = 1,\dots,n)$$

are n scalars and L_i $(i = 1,\dots,n)$ are n linear functionals of the following type :

$$(1)\ L_i(x) = \sum_{\substack{k=1,2,3\\ j=1,\dots,q}} \alpha_{ijk}\, x_k(t^j) + \beta_{ijk} \frac{\partial x_k}{\partial t_1}(t^j) + \gamma_{ijk} \frac{\partial x_k}{\partial t_2}(t^j)$$

with $x_k : T \longrightarrow \mathbb{R}$ $(k = 1,2,3)$ such that
$x(t) = (x_1(t), x_2(t), x_3(t))$,
$t^j (j = 1,\dots,q)$ are q points of T,

$(\alpha_{ijk}, \beta_{ijk}, \gamma_{ijk}) \in \mathbb{R}^3$ $(i = 1,\dots,n\ ;\ k = 1,2,3\ ;\ j = 1,\dots,q)$
are a set of scalars.

We present a solution to problem P_2 that generalizes the results presented by Duchon in [1], [2]. Our presentation follows the schema of [2].

II. PARAMETRIC SPLINE FUNCTIONS

II.1 NOTATIONS

Let m be an integer $m \geq 2$.
Let $X = D^{-m}L^2(\mathbb{R}^2)$, with semi-scalar product :

$$(u,v)_m = \sum_{i_1,\dots,i_m=1}^{2} \int_{\mathbb{R}^2} D_{i_1,\dots,i_m} u \ D_{i_1,\dots,i_m} v$$

The derivations are understood in the distribution sense.

$$|u|_m = (u,u)_m^{1/2} .$$

Let P_k be the space of polynomials whose degree does not exceed k.
The quotient $D^{-m}L^2/P_{m-1}$, with norm $\dot{v} \rightarrow |v|_m$ is known to be a Hilbert space.
The following result is due to J. Duchon.

PROPOSITION 1. Let L be a distribution with compact support of order $\leq$ m-2, satisfying $L(p) = 0, \forall p \in P_{m-1}$.

Then the mapping $\dot{v} \to L(v)$ from $D^{-m}L^2/P_{m-1}$ in $\mathbb{R}$ is definite and continuous.

Krylov's theorem can be used to prove that the distributions belonging to $D^{-m}L^2$ are in fact C^{m-2} functions.

Let p be a given integer ($p \geq 2$; p is the dimension of the space).

II.2 NOTATIONS IN A p-DIMENSIONAL SPACE

Let $Y = X^p$, with semi-scalar product

$$((y^1,y^2))_m = \sum_{i=1}^{p} (x_i^1, x_i^2)_m$$

with $y^i = (x_1^i,\dots,x_p^i) \quad i = 1,2$

$$\|y\|_m = ((y,y))_m^{1/2} .$$

Using the results stated in section 1-1, it is obvious that $(D^{-m}L^2)^p/(P_{m-1})^p$ with norm $\dot{v} \to \|v\|_m$ is a Hilbert space.

The associated canonical surjection is called π_m^p.

Let L be a linear continuous functional defined on Y, with compact support.

Let $L^k : Y \to \mathbb{R}$, be defined by

$$L^k(y) = L^k(x_1,\dots,x_p) = L(0,\dots,0, x_k,0,\dots,0)$$

Let L^Y be the mapping : $Y \to \mathbb{R}^p$, defined by

$$L^Y(y) = (L^1(y),\dots,L^p(y))$$

PROPOSITION 2. Let L be a linear continuous functional defined on Y with compact support, satisfying $L(q) = 0, \forall q \in (P_{m-1})^p$, then the mapping $\dot{v} \to L(v)$ from $(D^{-m}L^2)^p/(P_{m-1})^p$ in $\mathbb{R}$ is definite and continuous.

Proof. Let $v \in (D^{-m}L^2)^p \qquad v = (v_1,\dots,v_p)$

$$L(v) = \sum_{k=1}^{p} L^k(v) = \sum_{k=1}^{p} L^k(v_k)$$

(L^k denotes as well the mapping $(D^{-m}L^2)^p \to \mathbb{R}$ and the mapping $D^{-m}L^2 \to \mathbb{R}$ trivially asociated).

Now, $\forall k \in \{1,\dots,p\}$, L^k satisfies the assumptions of proposition 1.

So, for any k, the mapping $\dot{v}_k \longrightarrow L^k(v_k)$ is definite and continuous. Hence, the mapping $\dot{v} \longrightarrow L(v)$ is definite and continuous. □

II.3 THEOREM 1.

Let L_i $(i = 1,\ldots,n)$ be n given linear continuous functionals on Y, with compact support, whose partial mappings are distributions of order $\leq m-2$.
Let $z = (z_1,\ldots,z_n)$ be an element of $\mathbb{R}^n$.
Let $I = \{1,\ldots,n\}$
Assume that : $\exists I_o \subset \{1,\ldots,n\}$ such that :

$$\forall (\gamma_i)_{i=1,\ldots,n} \in \mathbb{R}^n,\ \exists ! q \in (P_{m-1})^p\ L_i(q) = \gamma_i, \quad \forall i \in I_o$$

Then there exists one and only one

$\sigma \in I_z = \{f \in (D^{-m}L^2)^p,\ \ L_i(f) = z_i,\ \forall i \in I\}$ with minimal semi-norm.

Proof. I_o can be associated with a basis $\{p_i,\ i \in I_o\}$ of $(P_{m-1})^p$, defined by $L_i(p_j) = \delta_{ij}$.
Any class $\dot{v}$ belonging to $(D^{-m}L^2)^p/(P_{m-1})^p$ possesses a unique representative w satisfying $L_i(w) = o\ \forall i \in I_o$. This w is given by $v - \sum_{i \in I_o} L_i(v)p_i$.
Any class $\dot{v}$ of $(D^{-m}L^2)^p/(P_{m-1})^p$ belongs to $\pi^p_m(I_z)$ if and only if

$$L_j(v - \sum_{i \in I_o} L_i(v)p_i) = L_j(\sigma - \sum_{i \in I_o} L_i(\sigma)p_i)\ \ \forall j \in I \backslash I_o.$$

This means that $\pi^p_m(I_z)$ is the intersection of the hyperplanes

$$\pi_j = \{\dot{v} \in (D^{-m}L^2)^p/(P_{m-1})^p\ ;\ L_j(v - \sum_{i \in I_o} L_i(\sigma)p_i) = L_j(\sigma - \sum_{i \in I_o} L_i(\sigma)p_i)\}$$

$\forall j \in I \backslash I_o$.

Each π_j is closed, since the functional $v \longrightarrow L_j(v - \sum_{i \in I_o} L_i(v)p_i)$ is

continuous on $(D^{-m}L^2)^p/(P_{m-1})^p$

(indeed, this functional vanishes on $(P_{m-1})^p$, since :

$\forall q \in (P_{m-1})^p$, $q(t) = \sum_{i \in I} p_i(t)L_i(q)$, because $(p_i)i \in I_o$ is a basis of $(P_{m-1})^p$.

Thus, the studied functional satisfies the assumptions of proposition 2 .

So $\pi_m^p(I_z)$ is a non empty closed space of $(D^{-m}L^2)^p/(P_{m-1})^P$.

So, it possesses a unique element with minimal norm, $\dot{s}$, whose intersection with I_z is

$$\sigma = s - \sum_{i\in I} L_i(s)p_i + \sum_{i\in I} z_i p_i \quad \square$$

II.4 CHARACTERIZATION

Let $\Delta_p^m : (D^{-m}L^2)^p \rightarrow (L^2)^p$

be defined by $\Delta_p^m(y) = (\Delta^m y_1, \ldots, \Delta^m y_p)$.

PROPOSITION 3.

$\Delta_p^m(\sigma)$ satisfies $\Delta_p^m(\sigma) = \sum_{i\in I} \lambda_i L_i^y$

$$\text{and} \quad \sum_{i\in I} \lambda_i L_i^y(q) = 0 \quad \forall q \in (P_{m-1})^p.$$

Proof. σ is the orthogonal projection of 0 into $\pi_m^p I_z$, so it is orthogonal to $\pi_m^p I_o$.

Hence $((\sigma,v))_m = 0$ for all $v \in (D^{-m}L^2)^p$.

Let use choose $\beta_i \in (D^{-m}L^2)^p$, such that $L_i(\beta_j) = \delta_{ij}$, then for all $v \in (D^{-m}L^2)^p$ for all $i \in I$, $L_i \; (v - \sum_{i\in I} L_j(v)\beta_j) = o$.

So $((\sigma,v))_m = \sum_{i\in I} (\sigma, \beta_j)L_j(v)$.

Define $\lambda_j = ((\sigma, \beta_j))_m$, for all $\varphi \in D^p$, $\varphi = (\varphi_1, \ldots, \varphi_p)$,

$((\sigma,\varphi))_m = \sum_{i\in I} \lambda_i L_i(\varphi)$ but

$$((\sigma, \varphi))_m = \sum_{k=1,\ldots,p} \sum_{i_1,\ldots,i_m=1}^{2} \int_{\mathbb{R}^2} D_{i_1}\ldots D_{i_m}\sigma_k \, D_{i_1}\ldots D_{i_m}\varphi_k.$$

So in the distribution sense :

$$((\sigma, \varphi))_m = \sum_{k=1,\ldots,p} (-1)^m \sum_{i_1,\ldots,i_m=1}^{2} \langle D_{i_1}^2 \ldots D_{i_m}^2 \sigma_k, \varphi_k \rangle$$

so for all $\varphi \in D^p$.

$$\sum_{k=1}^{p} \langle (-1)^m \Delta^m \sigma_k, \varphi_k \rangle = \sum_{i \in I} \langle L_i, \varphi \rangle$$

Define $\varphi = (o,o,\ldots,\varphi_k,o,\ldots,o)$, we conclude

$$\Delta^m \sigma_k = \sum_{i \in I} L_i^k \qquad \text{for } k = 1,\ldots,p$$

Therefore $\Delta_p^m \sigma = \sum_{i \in I} \lambda_i L_i^y$

Now using

$$((\sigma,v))_m = \sum_{i \in I} \lambda_i L_i(v) \text{ we conclude}$$

for all $v \in (P_{m-1})^p$ $\sum_{i \in I} \lambda_i L_i(v) = o$. □

PROPOSITION 4. Assume $g \in (D^{-m}L^2)^p$ and $\Delta_p^m g = \Delta_p^m \sigma$.

Then $g - \sigma \in (P_{m-1})^p$

Proof. Define $u = g - \sigma$, $u \in D^{-m}L^2$.

For all $k \in \{1,\ldots,p\}$ u_k is a tempered distribution satisfying $\Delta^m u_k = o$

Therefore (3) u is polynomial

Now its derivatives of order m belong to L^2. Therefore they vanish, so $u_k \in P_{m-1}$. Therefore $u \in (P_{m-1})^p$. □

PROPOSITION 5. Define $K_m(t) = |t|^{2m-2} \operatorname{Log} |t| \,/\, 2\pi(2^{m-1}(m-1)!)^2$.

Let L be a distribution on X, with compact support, orthogonal to P_{m-1}.

Then $L * K_m \in D^{-m}L^2$ and $\Delta^m(L * K_m) = L$

Proof. By Duchon [2] (this proposition needs not to be generalized for our purpose). □

DEFINITION. Let L be a distribution on Y, with compact support, orthogonal to $(P_{m-1})^p$. We define $L ** K_m \;:\; (D^{-m}(L^2))^p \to \mathbb{R}^p$

$$V \rightarrow (L^1 * K_m(V_1),\dots,L^p * K_m(v_p))$$

PROPOSITION 6. Let L be a distribution with a compact support, orthogonal to $(P_{m-1})^p$, then $L ** K_m \in (D^{-m}L^2)^p$ and $\Delta_p^m(L * * K_m) = L^y$.

Proof. $\Delta_p^m (L ** K_m) = (\Delta^m L^1 * K_m,\dots,\Delta^m L^p * K_m)$
Now the fact that L is orthogonal to $(P_{m-1})^p$ implies that L^k is orthogonal to P_{m-1}, $k = 1,\dots,p$. Therefore, using proposition 5,

$$\Delta_p^m (L * * K_m) = (L^1,\dots,L^p) = L^y. \square$$

THEOREM 2. σ can be written in an unique way as :

$$\sigma(t) = \sum_{i \in I} \lambda_i L_i ** K_m(t) + w(t)$$

with $w \in (P_{m-1})^p$ and $\sum_{i \in I} \lambda_i L_i(q) = o$ for all $q \in (P_{m-1})^p$.

Proof. proposition 3 implies that

$$\Delta_p^m(\sigma) = \sum_{i \in I} \lambda_i L_i^y$$

$$\sum_{i \in I} \lambda_i L_i(q) = o \quad \text{for all } q \in (P_{m-1})^p$$

Let $g = (\Delta_p^m \sigma) ** K_m = \sum_{i \in I} \lambda_i L_i^y * * K_m$

then $g \in (D^{-m}L^2)^p$ and satisfies (thanks to proposition 6)

$$\Delta_p^m g = \Delta_p^m \sigma$$

Therefore (proposition 4) $\sigma = g + w$, $w \in (P_{m-1})^p$. $\square$

III. APPLICATION OF THEOREM 2 TO THE SOLUTION OF PROBLEM P_2

In this case, the dimension of the space is $p = 3$.
To define the surface with linear functionals of type (1), we have to use a value of $m \geq 3$. We shall use $m = 3$.

In this case, the space Y is $Y = [D^{-3}L^2(\mathbb{R}^2)]^3$ and the solution of P_2 minimizes :

$$\|x\|^2 = \sum_{k=1}^{3} \sum_{i_1,i_2,i_3=1}^{2} \int_{\mathbb{R}^2} \left| \frac{\partial^3 x^k}{\partial t_{i_1} \partial t_{i_2} \partial t_{i_3}} \right|^2 dt$$

the kernel K_3 is defined by $K_3(t) = |t|^4 \text{ Log } (t)$

$L_i ** K_3$, for L_i of type (1) is defined by :

$$L_i ** K_3(t) = \sum_{k=1,2,3} \sum_{j=1}^{q} \alpha_{ijk} K_3(t-t^j) + \beta_{ijk} \frac{\partial K_3}{\partial t_1}(t-t^j) + \gamma_{ijk} \frac{\partial K_3}{\partial t_2}(t-t^j)$$

and the solution of P_2 is written :

$$\sigma(t) = \sum_{i=1}^{n} \lambda_i L_i ** K_3(t) + w(t)$$

where $(\lambda_i)_{i=1,\dots,n}$ and the coefficients of w(t) are solution of the linear system :

$$L_i(\sigma) = z_i \qquad i = 1,\dots,n$$

$$\sum_{i=1}^{n} \lambda_i L_i(g_j) = 0 \qquad j = 1,\dots,18 \left(\begin{array}{l} \text{where } (g_j)\ j=1,\dots,18 \text{ is} \\ \text{a basis of } (P_2)^3 \end{array} \right.$$

III. CONCLUSION

This method is presently experimented with using actual X-ray photographies and test surfaces. The first experiments tend to prove that the shape of the medullary canal is reconstructed with sufficient accuracy for orthopaedist surgery. In the automatic design of hip prosthis, the input to the parametric spline method will be the outline of the canal detected by image processing. The output will control the milling cutter tooling the prosthesis.

REFERENCES

1. DUCHON J. Interpolations des fonctions de deux variables suivant le principe de la flexion des plaques minces. R.A.I.R.O. Analyse Numérique Vol.10, n°12, 1976, pp.5-12

2. DUCHON J. Fonctions splines homogènes à plusieurs variables. Thèse d'Etat. Grenoble 1980.

3. SCHWARTZ L. Théorie des distributions. Hermann 1966.

4. CINQUIN P., CHALMOND B., BERARD D., DUSSERRE L., TROUILLOUD P.; GRAMMONT P., BINNERT D., MABILLE J.P., CARASSO C.. Hip prosthesis design.Medical informatics europe - DUBLIN - 22-25 mars 1982.

ON THE PENALTY METHOD FOR CONSTRAINED VARIATIONAL INEQUALITIES

J. Gwinner

Daimler-Benz AG
Stuttgart, West Germany

ABSTRACT

The penalty method of mathematical optimization is extended for the solution of constrained monotone variational inequalities. This method transforms the constrained v.i. into a sequence of simpler structured v.i. problems using a penalty functional that describes the constraints. Existence of solutions to these "penalized" auxiliary problems and their convergence to a solution of the original v.i. are proved. Then generalizing results of Bertsekas, Han and Mangasarian to infinite dimensions we present an exactness result, which permits to reduce the original problem to the solution of one single penalized v.i. . To this end we construct exact penalty functionals from the linear operator that describes the equality constraints and from the convex operator that describes the constraints of inequality type. A simplification of a penalized v.i. to a variational equality needs a differentiable penalty functional, however. In the case of finitely many inequality constraints we therefore consider a regularized penalty functional; here the associated penalized v.i. yields an approximate solution of the original v.i. . Finally we give an outlook over the application of this method to problems in continuum mechanics.

I. PROBLEM FORMULATION

Let E_x be a real linear normed space, X a closed, convex subset of E_x. Let f be a real-valued functional defined on $X \times X$ that is always supposed to satisfy for any x in X

$$f(x,x) = 0,$$

$f(x,.)$ is convex and lower semicontinuous on X.

In this contribution we are concerned with m o n o t o n e variational inequalities: f is monotone and hemicontinuous. We call f monotone, if for any x,y in X

$$f(y,x) + f(x,y) \leq 0$$

holds. For such functionals the following mild continuity condition with respect to the first variable is important: f is hemicontinuous, if for any y,u in E_x, $f(y+tu,y)$ is an upper semicontinuous function in the real t, where y+tu is restricted to vary in X only.

In addition, let us introduce explicit constraints in the now standard form of convex optimization. Let E_y, E_z be real linear normed spaces, let E_y be ordered by a convex, closed cone K in the following way

$$y_1 \geq y_2 \Leftrightarrow y_1 - y_2 \in K.$$

Let h, resp. g denote an affine-linear mapping on X with values in E_z, resp. a K-convex (i.e., convex with respect to the ordering above) mapping on X with values in E_y. Then we consider the "feasible" set S, given by

$$S = \{x \in X: g(x) \leq 0_y, \ h(x) = 0_z\}.$$

S is clearly a convex set; it is always supposed to be closed, what can be guaranteed by appropriate continuity assumptions on g and h.

Now the problem of the constrained variational inequality defined by f, g, h, and X consists in finding some $\bar{x}$ in S such that

$$f(\bar{x},x) \geq 0$$

is valid for any x in S, provided the solution set of this problem (f,S)

$$L(f,s) = \{\bar{x} \in S: f(\bar{x},x) \geq 0 \ \forall x \in S\}$$

is not void.

The problem of constrained monotone variational inequalities subsumes constrained convex optimization problems, constrained convex-concave minimax problems, and constrained monotone operator inequalities. Indeed, if j is a convex lower semicontinuous functional on X, consider

$$f(x,y) = j(y) - j(x) \qquad x,y \in X,$$

and with S given above, we arrive at a constrained convex optimization problem. If j is more generally a concave-convex upper-lower semicontinuous functional on $X^1 \times X^2$ with feasible sets

$$S^i = \{x_i \in X^i: \ g_i(x_i) \le 0, \ h_i(x_i) = 0\} \qquad (i=1,2),$$

then the problem of finding a saddle point $(\bar{x}_1,\bar{x}_2)$ in $S^1 \times S^2$, i.e.,

$$j(x_1,\bar{x}_2) \le j(\bar{x}_1,\bar{x}_2) \le j(\bar{x}_1,x_2) \qquad \forall x_1 \in S^1, \forall x_2 \in S^2$$

is equivalent to a constrained monotone variational inequality with the following data:

$$S = S^1 \times S^2, \qquad X = X^1 \times X^2,$$

$$f(x,y) = j(x_1,y_2) - j(y_1,x_2), \quad x = (x_1,x_2), \ y = (y_1,y_2) \in X,$$

$$g = (g_1,g_2), \ h = (h_1,h_2).$$

Monotone operator inequalities on the subset S can also be included in the subsequent way. Let m be an operator on X with values in E_x', the topological dual of E_x. Obviously the operator m is monotone, if and only if

$$f(x,y) = (m(x),y-x) \qquad x,y \in X$$

is monotone. To treat multivalued operators m with convex, $\sigma(E_x',E_x)$-compact values, we modify f to

$$f(x,y) = \max \{(x',y-x): \ x' \in m(x)\}.$$

Then it can be proved by a strong separation argument (see p. 5 in [13]) or by the application of a fixed point theorem (see Prop. 7.1 in [5]) that $\bar{x}$ in S solves the associated monotone variational inequality, if and only if some $\bar{x}'$ in $m(\bar{x})$ exists such that for all y in S

$$(\bar{x}', \ y-\bar{x}) \ge 0$$

holds.

II. PRELIMINARIES AND A BASIC EXISTENCE THEOREM

In this section, let f be defined on $C \times C$, with C convex and closed. A useful tool for the study of monotone variational inequalities is the subsequent

PROPOSITION 2.1. Let D be a convex subset of C. Let f be monotone and hemicontinuous. Then the two following statements are equivalent:

(i) $x \in D,\ f(x,y) \geq 0 \quad \forall y \in D;$

(ii) $x \in D,\ f(y,x) \leq 0 \quad \forall y \in D.$

The proof of (i) $\Rightarrow$ (ii) is obvious by monotonicity; (ii) $\Rightarrow$ (i) follows from a limit argument using convexity and hemicontinuity (for details see e.g. [12] p. 375).

PROPOSITION 2.2. Let in addition g be a convex, lower semicontinuous functional, defined on the Banach space E_x. Let ∂g denote the subdifferential of g. Then the subsequent assertions are equivalent:

(i) $x \in C,\ f(x,y) + g(y) - g(x) \geq 0 \quad \forall y \in C;$

(ii) $x \in C,\ f(x,y) + \max\{(x',y-x) : x' \in \partial g(x)\} \geq 0 \ \forall y \in C;$

(iii) $x \in C,\ x' \in \partial g(x),\ f(x,y) + (x',y-x) \geq 0 \quad \forall y \in C.$

For its proof we notice, since $f(x,x) = 0$, that (i) means: x minimizes the convex functional $f(x,.) + g$ on C. Therefore Proposition 4.1 of [12] immediately yields the claimed equivalence of the assertions above.

THEOREM 2.3. Let f be a monotone, hemicontinuous functional on $C \times C$, where C is a closed, convex subset of the reflexive Banach space E_x. If (a) C is bounded, or if (b) f is coercive, i.e.,

$$f(x,x_o)\ /\ \|x\| \to -\infty \qquad (x \in C,\ \|x\| \to \infty)$$

holds with some x_o in C, then there exists a solution $\bar{x}$ in L(f,C).

For the proof of this existence theorem we refer to [12] (p. 375-377). This existence result for monotone variational inequalities can be established in a more general setting (see e.g. Theorem 1 in [18], Satz 1.8 in [13]).

The functional f on $C \times C$ is called <u>uniformly monotone</u>, if there exists some positive constant a such that for any x_1, x_2 in C

$$f(x_1,x_2) + f(x_2,x_1) \leq -a\,\|x_1-x_2\|^2$$

holds. A convex functional j on C is called <u>uniformly convex</u>, if there exists some positive constant b such that for any x, y in C

$$j\left(\frac{x+y}{2}\right) \leq \frac{1}{2}\,j(x) + \frac{1}{2}\,j(y) - b\|x-y\|^2$$

holds. With these notions we have

LEMMA 2.4. (1) Let f be uniformly monotone. Let x_o, x_ε ($\varepsilon>0$) be given in C such that for any x in C

$$f(x_o,x) \geq 0, \qquad f(x_\varepsilon,x) \geq -\varepsilon$$

is valid. Then

$$\| x_\varepsilon - x_o \|^2 \leq \varepsilon a^{-1}.$$

(2) Let j be uniformly convex. Let x_o, x_ε ($\varepsilon>0$) be given in C such that for any x in C

$$j(x_o) \leq j(x), \qquad j(x_\varepsilon) \leq j(x) + \varepsilon$$

is valid. Then

$$\| x_\varepsilon - x_o \|^2 \leq \frac{1}{2}\varepsilon b^{-1}.$$

Proof. (1) Choose $x = x_\varepsilon$, resp. $x = x_o$ and add up the two inequalities. (2) Choose $x = \frac{1}{2}(x_\varepsilon + x_o)$ in the two inequalities and add up. □

We remark that the global requirements of uniform monotonicity, resp. uniform convexity can be easily weakened to local assumptions for elements of the respective solution set (cf. [12] p. 377, resp. p. 381).

III. THE PENALTY METHOD FOR CONSTRAINED VARIATIONAL INEQUALITIES

In this section we extend the penalty method of mathematical optimization, originally introduced by Courant [7], to the solution of constrained variational inequalities, given by a monotone functional f and a nonvoid closed convex proper subset S of X. This method transforms the constrained problem on S into a simpler structured variational inequality on X only by means of a penalty functional p. We suppose that the real-valued functional p on X satisfies

$$\text{(HP)} \quad \begin{cases} \text{p is convex and lower semicontinuous} \\ \text{p is nonnegative} \\ x \in X \;\&\; p(x) = 0 \iff x \in S. \end{cases}$$

Then we can consider the "penalized" variational inequality with positive penalty parameter r

$$\text{(3.1)} \quad x^r \in X, \quad f(x^r,y) + r(p(y) - p(x^r)) \geq 0 \qquad \forall\, y \in X.$$

If $X = E_x$, $f(x,y) = (Tx, y-x)$, T monotone operator (very often in applications), in view of Proposition 2.2, (3.1) is equivalent to the inclusion

$$0 \in Tx^r + r\ \partial p(x^r),$$

and if in addition p is differentiable, then (3.1) reduces to the variational equality

$$Tx^r + r\ \nabla p(x^r) = 0.$$

The essential properties of this method: existence of the approximations x^r and convergence to a solution of the original problem (f,S) are summarized in the following

THEOREM 3.1. Suppose, f is coercive with some x_o in S, contained in the reflexive Banach space E_x. Then solutions x^r exist for any nonnegative r. Every sequence (x^i), generated by $r = r_i$, $r_i \to \infty$, possesses weak limit points, and every weak limit point of (x^i) is a solution of (f,S).

Proof. By coerciveness of f, we obtain

$$(3.2) \quad -f^r(x,x_o) / \| x \| \geq -f(x,x_o) / \| x \| \to \infty$$

for x in X, $\| x \| \to \infty$, where f^r is given by

$$f^r(x,y) = f(x,y) + r(p(y) - p(x)).$$

Since f^r is monotone and hemicontinuous on X, Theorem 2.3 applies. It yields the existence of solutions x^r to (f^r,X). Now for any x_s in S

$$(3.3) \quad f(x^r,x_s) \geq 0$$

is valid. In view of (3.2) and (3.3), every sequence (x^i) for $r = r_i \to \infty$ is bounded, hence weak limit points $\overline{x}$ of (x^i) exist. Since by monotony

$$p(x^i) \leq -\frac{1}{r_i} f(x_s,x^i)$$

and by weak lower semicontinuity

$$p(\overline{x}) \leq \liminf p(x^i) \leq -\liminf \frac{1}{r_i} f(x_s,x^i) = 0,$$

$\overline{x}$ belongs to S. From (3.3) it follows by monotony and weak lower semicontinuity that for any x_s in S

$$f(x_s,\overline{x}) \leq \liminf f(x_s,x^i) \leq 0.$$

In virtue of Proposition 2.1, $\overline{x}$ lies in L(f,S). □

Remarks. If the solution of (f,S) is unique, then the entire sequence (x^i) converges weakly to its solution. This uniqueness and in addition, strong convergence can be easily shown (apply Lemma 2.4) under the hypothesis of uniform monotonicity. Here we can refer to Theorem 2.1 in [12], for penalty methods and regularization are strongly related. Indeed, (3.1) can be formulated as

$$x^r \in X, \quad (p(y) - p(x^r)) + \frac{1}{r} f(x^r,y) \geq 0 \quad \forall y \in X$$

and $f \equiv \rho$ can be interpreted as a regularizing monotone functional for the special variational inequality associated to the minimization of p on X having S as its solution set.

Since we are interested in feasible sets S that are given by

$$S = \{x \in X: \ g(x) \leq 0, \ h(x) = 0\},$$

the question arises how the penalty functional p can be constructed from the operators g and h. As for the equality constraint with a linear-affine continuous operator h, we can simply choose

$$p(x) = \| h(x) \|_z$$

(or some monotone convex function of it). For the generalized inequality constraint $g(x) \in -K$, however, we must insert an operator k which returns the deviation of g(x) from the set -K in the space E_y such that $\| k \circ g(\cdot) \|_y$ satisfies (HP). In the next section we list some examples of appropriate operators k.

Let us now investigate the exactness of the penalty method for constrained variational inequalities presented in this paper. Exactness means that for a sufficiently large penalty parameter r the solution sets of the penalized problem and of the original constrained problem coincide. Here we extend a result due to Bertsekas (Proposition 1 in [4]), Han and Mangasarian (Theorem 4.9 in [14]) to our more general setting. To this end we need the concept of a Lagrange-Kuhn-Tucker multiplier $(\bar{y}',\bar{z}')$ associated to some fixed $\bar{x}$ in L(f,S). Note that $\bar{x}$ in L(f,S), since $f(\bar{x},\bar{x}) = 0$, in particular solves the subsequent convex optimization problem

$$\text{minimize } f(\bar{x},x) \text{ subject to } x \in X, \ g(x) \leq 0, \ h(x) = 0.$$

Therefore a pair $(\bar{y}',\bar{z}')$ in the dual space $E_y' \times E_z'$ with $\bar{y}' \geq 0$ (i.e. $\bar{y}'$ is in the dual cone $K' = K^+$) is called a <u>Lagrange-Kuhn-Tucker multiplier</u> associated to $\bar{x}$, if for any x in X

(LKT) $\quad 0 \le f(\bar{x},x) + (\bar{y}',g(x)) + (\bar{z}',h(x))$

holds. Existence assertions for Lagrange-Kuhn-Tucker multipliers are provided by the now well-known theory of convex optimization.
The penalty functional p is now given by

$$(3.4) \quad p(x) = Q(\| k(g(x)),h(x) \|).$$

It involves a norm $\|\cdot\|$ in $E_y \times E_z$, a continuous monotone convex function Q with $Q(0) = 0$ and $Q(t) > 0$ for $t > 0$, and a continuous operator $k: E_y \to E_y$ that vanishes on $-K$. Here k is called <u>K-isotone</u>, if $k(y) \ge y$ (in the ordering given by K) holds for any y in E_y.

PROPOSITION 3.2. Let k be K-isotone. Suppose that the function Q satisfies

$$Q'(0+) = \lim_{t \to 0+} \frac{1}{t} Q(t) > 0.$$

Let $(\bar{y}',\bar{z}')$ be some Lagrange-Kuhn-Tucker multiplier associated to $\bar{x}$ in L(f,S). Let

$$\bar{r} = Q'(0+)^{-1} \| \bar{y}',\bar{z}' \|',$$

where $\|\cdot\|'$ denotes the dual norm to some norm in $E_y \times E_z$. Then for $r \ge \bar{r}$, $\bar{x}$ solves the penalized problem (f^r,X); for $r > \bar{r}$, every solution x^r to (f^r,X) solves the original problem (f,S).

Proof. Essentially, we modify the arguments of Bertsekas and of Han and Mangasarian. Let $r \ge \bar{r}$. Then we have for any x in X

$$
\begin{aligned}
(3.5) \quad f^r(\bar{x},x) &= f(\bar{x},x) + r\, Q(\| k(g(x)),h(x) \|) && (\text{since } p(\bar{x}) = 0) \\
&\ge f(\bar{x},x) + \frac{\| \bar{y}',\bar{z}' \|'}{Q'(0+)} Q(\| k(g(x)),h(x) \|) \\
&\ge f(\bar{x},x) + \| \bar{y}',\bar{z}' \|' \, \| k(g(x)),h(x) \| && (\text{since Q is convex}) \\
&\ge f(\bar{x},x) + (\bar{y}',k(g(x))) + (\bar{z}',h(x)) \\
&\ge f(\bar{x},x) + (\bar{y}',g(x)) + (\bar{z}',h(x)) && (\text{k is K-isotone}) \\
&\ge f(\bar{x},\bar{x}) = 0 && (\text{by (LKT)}).
\end{aligned}
$$

Therefore $\bar{x}$ belongs to $L(f^r,X)$. To prove the second part, we only have to verify that x^r lies in the feasible set S. Assume not. Then $h(x^r) \neq 0$, or $g(x^r)$ does not belong to $-K$, hence $k(g(x^r)) \neq 0$, since k is K-isotone. In any case, the second row of (3.5) holds for $x = x^r$ with a strict inequality. Therefore the estimates above imply that

$$f^r(\bar{x},x^r) > 0,$$

a contradiction in view of Proposition 2.1 applied to f^r on X. □

IV. SOME PENALTY FUNCTIONALS FOR INEQUALITY CONSTRAINTS

Here we give some examples for the operator k (as the interesting part of the penalty functional in (3.4)) that satisfy all the assumptions demanded in the preceding section. Let us first list some examples for the case of a finite number of inequality constraints ($E_y = \mathbb{R}^M$, $K = \mathbb{R}^M_+$):

(4.1) $k_m(y_m) = \max\ (0,y_m)$ (Han and Mangasarian [14])

(4.2) $k_m(y_m) = (\max(0,y_m) + 1)^{n_m} - 1$ with positive integers n_m

(4.3) $k_m(y_m) = \max\ (0,\ -1 + \exp(c_m y_m))$ with $c_m > 0$ (Evans and Gould [9])

(4.4) $k_m(y_m) \begin{cases} = 0 & y_m \leq 0 \\ > 0 & y_m > 0 \end{cases}$ k_m convex, continuous with $k_m'(0+) > 0$ (Bertsekas [4])

where $k(y) = (k_m(y_m))_{m=1}^M$. Here any l_q-norm ($q \geq 1$) or the l_∞-norm can be chosen; in the known penalty methods for optimization problems, mostly the l_1-norm is taken.

The example (4.4) is covered by Proposition 3.2, since with

$$c := \min_m k_m'(0+) > 0$$

we can write

$$Q(\|(k_m(y_m))\|) = Q(c\|(c^{-1}k_m(y_m))\|) =: \tilde{Q}(\|(\tilde{k}_m(y_m))\|)$$

and $\tilde{k}_m = c^{-1}\ k_m$ is K-isotone by convexity. Evidently all the examples above give rise to convex penalty functionals

$$p(x) = Q(\|(k_m(g_m(x)))_{m=1}^M\|),$$

provided all the g_m ($m=1,\ldots,M$) are convex.

As a way to extend (4.1) to infinite dimensions to treat constrained optimization problems by infinite penalty methods in a Hilbert space, Wierzbicki and Kurcyusz [25] studied the projection on a cone. More specifically, let the dual cone K' be defined by

$$K' = \{v \in E_y : (v,u) \geq 0 \text{ for all } u \text{ in } K\}$$

in the Hilbert space E_y, endowed with the scalar product $(.,.)$. Let the mapping k' assign to every y in E_y

$$k'(y) = \mathrm{proj}_{K'} y,$$

the unique projection of y onto the closed convex cone K'. Then k' is continuous and can be characterized by the following (see Lemma 2.2 in [25]).

LEMMA 4.1 For any y in E_y, y' in K', the equality $y' = k'(y)$ is equivalent to the two statements:

(i) $y'-y \in K$,

(ii) $(y',y'-y) = 0$.

This Lemma immediately implies that y belongs to $-K$, if and only if $k'(y) = 0$. Moreover, (i) means that k' is K-isotone.
In virtue of the decomposition theorem in [25] (Theorem 2.4) we can also write

$$k'(y) = y - \mathrm{proj}_{-K} y;$$

thus k' returns the deviation of y from the constraint $y \leq 0_y$. Since k' is positively-homogeneous (see Lemma 2.3 in [25]) and by Lemma 2.5 in [25]

$$\| k'(y_1+y_2-u) \| \leq \| k'(y_1) + k'(y_2) \|, \quad y_1, y_2 \in E_y$$

holds for any u in K, the penalty functional p, given now by

$$p(x) = Q(\| k'(g(x)) \|),$$

is convex, for we assumed g to be convex with respect to the ordering in E_y.

There is another way to extend (4.1) to infinite dimensions, provided the space E_y can be endowed with a lattice structure. We think here of spaces of real-valued functions that are defined on some open bounded set G in $\mathbb{R}^N$. Namely $C(G)$, the space of continuous functions, all spaces $L_q(G)$ and hence all Sobolev spaces $W_q^1(G)$ $(1 \leq q \leq \infty)$ (see also p. 219 in [15]) possess the property that with y, also

$$y^+ = \frac{1}{2}(y + |y|) = \sup(0,y)$$

lies in the respective space. Therefore to treat inequalities

$$y \leq 0 \iff y(s) \leq 0 \qquad \text{a.e. in } G,$$

we can consider

$$\bar{k}(y) = \sup\ (0,y).$$

Evidently $\bar{k}$ is K-isotone with respect to the cone K of nonnegative functions on G. Moreover, with any L_q-norm $(1 \le q \le \infty)$,

$$p(x) = Q(\| \bar{k}(g(x)) \|_q)$$

is convex because of convexity of g with respect to K, as can be seen as follows. Note, $\bar{k}$ is positively-homogeneous. Thus we have to verify for an arbitrary u in K

$$\| \bar{k}(y_1+y_2-u) \|_q \le \| \bar{k}(y_1) \|_q + \| \bar{k}(y_2) \|_q .$$

This inequality follows from

$$\max\ (0,t_1+t_2-t) \le \max\ (0,t_1) + \max\ (0,t_2),$$

where t is any nonnegative real, from the isotony of the L_q-norm for nonnegative functions:

$$0 \le y_1 \le y_2 \quad \text{a.e. in } G \Rightarrow \| y_1 \|_q \le \| y_2 \|_q ,$$

and finally from the Minkowski inequality.

An important drawback of the proposed penalty functionals above is their nondifferentiability. Thus the penalized variational inequalities cannot be simplified to variational equalities by means of Proposition 1.2. But differentiability cannot be expected with e x a c t penalty functionals. By his study of necessary conditions for exactness, Bertsekas [4] already brought this fact to attention and discussed it in detail for mathematical programming problems. A way out is to regularize the functionals above and investigate what remains from exactness. This will be done for the case of finitely many constraints $(E_y = \mathbb{R}^M,\ K = \mathbb{R}^M_+)$.

Instead of (4.1) let us consider for some $\varepsilon > 0$

$$1(y) = \left(\sqrt{\varepsilon^2 + y_m \max\ (0,y_m)} - \varepsilon\right)_{m=1}^{M},$$

and with the constraint functional $g = (g_1,\dots,g_M)$ the associated penalty functional reads now

$$p(x) = \| 1(g(x)) \|_1 \qquad\qquad x \in X.$$

Then p vanishes on S, is positive outside of S. Furthermore p inherits convexity and lower semicontinuity from the $g_m (m = 1,\dots,M)$. In addition, 1 is differentiable, but not K-isotone. We can modify the arguments of the proof of Proposition 3.2 in the following way:

Let $\bar{y}'$ be some Lagrange-Kuhn-Tucker multiplier associated to $\bar{x}$ in L(f,S).

Let

$$r \geq \| \bar{y}' \|_\infty = \max y_m .$$

Then for any x in X we have

$$f^r(\bar{x},x) = f(\bar{x},x) + r\|1\ (g(x))\|_1$$

$$\geq f(\bar{x},x) + \| \bar{y}' \|_\infty \| 1(g(x)) \|_1$$

$$\geq f(\bar{x},x) + (\bar{y}',1(g(x))),$$

and, since $(1(y))_m \geq y_m - \varepsilon$,

$$f^r(\bar{x},x) \geq f(\bar{x},x) + (\bar{y}',g(x)) - \varepsilon\| \bar{y}' \|_1$$

$$\geq -\varepsilon \|\bar{y}'\|_1 .$$

Thus $\bar{x}$ solves the penalized problem (f^r,X) "almost". Since f^r inherits uniform monotonicity from f, Lemma 2.4 applies to (f^r,X). Therefore $\bar{x}$ is indeed an approximate solution of this penalized problem under the additional hypothesis of uniform monotonicity, resp. under the hypothesis of uniform convexity in case of a convex optimization problem. Consequently, with these assumptions, an arbitrarily close approximation x^r to the solution $\bar{x}$ of the original problem can be obtained as the solution of a single penalized problem for a sufficiently large penalty parameter r and an appropriate small ε in the functional 1.

V. CONSTRAINED VARIATIONAL INEQUALITIES IN CONTINUUM MECHANICS AND THE PENALTY METHOD - AN OUTLOOK

The variational principles in continuum mechanics lead to variational boundary-value problems with partial differential operators where in the linear case and in a large class of nonlinear problems these differential operators are monotone. Thus to describe mathematically the deformation of a body under external forces monotone variational equalities or more generally monotone variational inequalities, when constraints are present, are a convenient tool. This variational approach is not only important from a theoretical point of view (existence theory), but also for the numerical solution; in particular the finite element solution can be seen as the solution of the variational problem in the finite dimensional subspace obtained by finite element discretization (see Oden & Reddy [20] for more details).

Our concept of monotone variational inequalities does not only include variational problems with monotone operators, but also convex optimization - useful if the involved operators are potential -, and convex-concave minimax problems - useful for the application of complementary-dual variational principles which are relevant for the formulation of the mixed finite element method in contrast to the more usual displacement finite element method.

Constraints can arise from material behaviour or from geometrical boundary conditions. The linear equality-type constraint of incompressibility is encountered for rubber-like materials and in fluid mechanics (see e.g. Prager [23], Temam [24]). Convex inequality-type contraints occur for instance in unilateral problems such as the unilateral contact problem or in elastoplastic analysis (see the monograph of Duvaut & Lions [8], see also Bufler [6], Nečas [19], Panagiotopoulos [22]).

The application of penalty methods to constrained problems in mechanics is of course not new; it can be traced back to Courant [7]. In the seventies the constraint of incompressibility in fluid mechanics has been treated by a special penalty-regularization method (see Temam [24] p. 426 ff; Girault & Raviart [11] p. 45 ff). The boundary constraint of Dirichlet, resp. of Neumann type was studied by Babuska [2], resp. Babuska & Zlamal [3]. Other applications can be found with Zienkiewicz and his co-workers [26-28]. Only recently Malkus [17] deals with the constraint of incompressibility in nonlinear elastics, Kikuchi & Song [16] derive convergence of a penalty method for unilateral problems with elastic bodies. In connection with the mixed finite element method, Oden [21] presents a general theory of penalty methods for nonlinear not necessarily monotone problems in continuum mechanics. In all these contributions, however, an unlimited growth of the penalty parameter is assumed (for details of these methods see also Auslender [1] p. 47 ff; Glowinski, Lions & Trémolières [10] p. 19, 44).

Real-world problems in engineering lead to a large number of degrees of freedom (=dimension of the finite dimensional subspace in the finite element approximation). Moreover if material nonlinearities and/or geometrical nonlinearities (large displacements, large strains) cannot be neglected, then the application of the standard finite element method - due to its inherent linearizations and approximations - is only feasible with an incremental solution strategy and with equilibrium iterations in each loading step. Therefore an additional infinite penalty method is not practicable. This explains our strong interest in finite penalty methods.

Penalty methods can be applied at two levels. First at a continuous level in an infinite dimensional setting - afterwards a discretization of the penalized problem by e.g. finite elements is performed -, or second after the discretization in finite dimensional problems. The resulting methods can be different. This gives a reason for the general infinite dimensional framework employed in this paper.

REFERENCES

1. A. Auslender, Problèmes de minimax via l'analyse convexe et les inégalités variationnelles. Théorie et algorithmes. Lect. Notes Econom. Math. Syst., 77. Springer, 1972.

2. I. Babuska, The finite element method with penalty. Math. Comput. 27 (1973), 221-228.

3. I. Babuska & M. Zlámal, Nonconforming elements in the finite element method with penalty. SIAM J. Numer. Anal. 10 (1973), 863-875.

4. D.P. Bertsekas, Necessary and sufficient conditions for a penalty method to be exact. Math. Programming 9 (1975), 87-99.

5. F.E. Browder, Nonlinear operators and nonlinear equations of evolution in Banach spaces. Proc. Symposia in Pure Math., Vol. 18,2. American Math. Soc., Providence, 1976.

6. H. Bufler, Zur Herleitung der Variationsungleichungen des reibungsfreien Kontaktproblems. GAMM-Tagung 1981.

7. R. Courant, Variational methods for the solution of problems of equilibrium and vibrations. Bull. Amerc. Math. Soc. 49 (1943), 1-23.

8. G. Duvaut & J.L. Lions, Les inéquations en mécanique et en physique. Dunod, Paris, 1972.

9. J.P. Evans & F.J. Gould, An existence theorem for penalty function theory. SIAM J. Control 12 (1974), 509-516.

10. R. Glowinski, J.L. Lions & R. Trémolières, Analyse numerique des inéquations variationnelles. Tome 1. Dunod, Paris, 1976.

11. V. Girault & P.-A. Raviart, Finite element approximation of the Navier-Stokes equations. Lect. Notes Math., 749. Springer, 1981.

12. J. Gwinner, On the regularization of monotone variational inequalities. Operations Research Verfahren 28 (1978), 374-388.

13. J. Gwinner, Nichtlineare Variationsungleichungen mit Anwendungen, Dissertation, Universität Mannheim. Haag & Herchen, Frankfurt, 1978.

14. S.P. Han & O.L. Mangasarian, Exact penalty functions. Math. Programming 17 (1979), 251-269.

15. B. Hanouzet & J.L. Joly, Méthodes d'ordre dans l'interprétation de certaines inéquations variationnelles et applications. J. Functional Analysis 34 (1979), 217-249.

16. N. Kikuchi & Y.J. Song, Remarks on relations between penalty and mixed finite element methods for a class of variational inequalities. Internat. J. Numer. Methods Engrg. 15 (1980), 1557-1579.

17. D.S. Malkus, Finite elements with penalties in nonlinear elasticity. Internat. J. Numer. Methods Engrg. 16 (1980), 121-136.

18. G.J. Minty, On variational inequalities for monotone operators, I. Advances Math. 30 (1978), 1-7.

19. J. Nečas, Variational inequalities in elasticity and plasticity with application to Signorini's problem and to flow theory of plasticity. ZAMM 60 (1980), T20-T26.

20. J.T. Oden & J.N. Reddy, Variational methods in theoretical mechanics. Springer, 1976.

21. J.T. Oden, A theory of penalty methods for finite element approximations of highly nonlinear problems in continuum mechanics. Computers & Structures 8 (1978), 445-449.

22. P.D. Panagiotopoulos, Convex analysis and unilateral static problems. Ingenieur-Archiv 45 (1976), 53-68.

23. W. Prager, Einführung in die Kontinuumsmechanik. Birkhäuser, Basel, 1961.

24. R. Temam, Navier-Stokes equations. Studies in Math. and its Appl., Vol. 2. North-Holland, Amsterdam, 1977.

25. A.P. Wierczbicki & S. Kurcyusz, Projections on a cone, penalty functionals and duality theory for problems with inequality constraints in Hilbert space. SIAM J. Control 12 (1974), 509-516.

26. O.C. Zienkiewicz, Constrained variational principles and penalty function methods in finite element analysis. In: Conference on the numerical solution of differential equations (G.A. Watson; ed), Lect. Notes Math., 363. Springer, 1963; pp. 207-214.

27. O.C. Zienkiewicz & P.N. Godbole, A penalty function approach to problems of plastic flow of metals with large surface deformations. J. Strain Analysis 10 (1975), 180-183.

28. O.C. Zienkiewicz & J.C. Heinrich, A unified treatment of steady-state shallow water and two-dimensional Navier-Stokes equations - finite element penalty function approach. FENOMECH' 78, Part III. Comput. Methods Appl. Mech. Engrg. 17/18 (1979), 673-698.

BANG-BANG-CONTROLS FOR TIME-OPTIMAL PARABOLIC BOUNDARY CONTROL PROBLEMS WITH INTEGRAL STATE CONSTRAINTS

Uwe Mackenroth

Universität Bayreuth
Bayreuth, West Germany

I. INTRODUCTION

Let $\Omega \subset \mathbb{R}^n$ be open and bounded. We assume that the boundary Γ of Ω is a C^∞-manifold of dimension n-1. Locally, Ω is totally on one side of Γ. With $T > 0$ put $Q_T :=]0,T[\times \Omega$, $\Sigma_T :=]0,T[\times \Gamma$. For $i,j = 1,\dots,n$ let be given functions a_o, $a_{ij} \in C^\infty(\bar{\Omega})$ and a real number $c > 0$ such that

$$a_{ij} = a_{ji} \qquad i,j = 1,\dots,n,$$

$$\sum_{i,j=1}^{n} a_{ij}(x)\xi_i \xi_j \geq c \sum_{i=1}^{n} \xi_i^2 \qquad \forall\, \xi \in \mathbb{R}^n, \forall\, x \in \Omega.$$

For $\varphi \in L^2(Q_T)$ define

$$A\varphi := -\sum_{i,j=1}^{n} \frac{\partial}{\partial x_i}\left(a_{ij}\frac{\partial \varphi}{\partial x_j}\right) + a_o\varphi.$$

Furthermore, let K be a convex subset of $C(\bar{\Omega})$ having a nonempty interior and let

$$f : [0,+\infty[\times \bar{\Omega} \times \mathbb{R} \longrightarrow \mathbb{R}$$

be a continuous function such that $f(t,x,\cdot)$ is convex for every $(t,x) \in [0,+\infty[\times \bar{\Omega}$. Finally, assume that α, m are positive real constants.

We introduce now the following time-optimal control problem.

(P) Find the smallest number $T > 0$ with the property that there exists a control $u \in L^\infty(\Sigma_T)$ with

$|u| \leqslant m$

such that the solution y of

(1.1) $\frac{\partial y}{\partial t} + A y = 0,$

(1.2) $\alpha y_{|\Sigma_T} + \frac{\partial y}{\partial n_A} = u,$

(1.3) $y(0) = 0$

fulfils

(1.4) $\int_\Omega f(t,x,y(t,x))\, dx \leqslant 0 \qquad \forall\, t \in [0,T],$

(1.5) $y(T) \in K.$

Of course, the solution y of (1.1) - (1.3) has to be understood in certain generalized sense (compare e.g. Lions [1], Chap. III). In Mackenroth [3], Theorem 1 (d), it is shown that y is continuous on $\bar{Q}_T$. Hence the constraints (1.4), (1.5) make sense.

More general time-optimal parabolic boundary control problems with state constraints are considered in Mackenroth [4]. Here we investigate more closely the case where the state constraint is given in integral form. Especially, we refine the results of Mackenroth [4] on bang-bang-properties of the optimal controls. This leads also to a new explanation and a generalization of a Theorem in Tröltzsch [8] on bang-bang-controls.

II. NECESSARY OPTIMALITY CONDITIONS

We often denote the solution of (1.1) - (1.3) belonging to $u \in L^\infty(\Sigma_T)$ by y(u). The following Theorem is an immediate consequence of Theorem 1 of Mackenroth [4].

THEOREM 1. Suppose that there exists $T > 0$ and a control $u \in L^\infty(\Sigma_T)$ with $|u| \leqslant m$ such that y(u) fulfils the constraints (1.4), (1.5). Moreover, assume

$$K = \mathrm{cl}_{L^2(\Omega)}(K) \cap C(\bar{\Omega}).$$

Then the time-optimal problem (P) has an optimal solution.

Next we want to derive necessary optimality conditions for (P). Therefore we need the adjoint partial differential equation. Let

$$B_T := \{\varphi \in H^{2,1}(Q_T) \mid \frac{\partial\varphi}{\partial t} + A\varphi \in \mathrm{cl}_{C(\bar{Q}_T)} (\mathcal{D}(Q_T)), \alpha\varphi_{|\Sigma_T} + \frac{\partial\varphi}{\partial n_A} = 0,\ \varphi(0) = 0\}.$$

For any $z^* \in C(\bar{Q}_T)^*$, $h^* \in C(\bar{\Omega})^*$ there exists a unique $w^* \in L^1(Q_T)$ with

$$(2.1)\quad \langle\frac{\partial\varphi}{\partial t} + A\varphi, w^*\rangle_{L^2(Q_T)} = \langle\varphi, z^*\rangle + \langle\varphi(T), h^*\rangle \qquad \forall\, \varphi \in B_T$$

(compare Mackenroth [3], Lemma 2 and Theorem 3). The solution w^* of (2.1) may be restricted to Σ_T in a certain generalized sense and $w^*_{|\Sigma_T}$ is summable (see Mackenroth [3], Sect. 3 and Theorem 4).

For a Banach space Y we denote by $NBV(0,T;Y)$ the space of functions $v : [0,T] \longrightarrow Y$ of bounded variation, which are continuous on the right on $]0,T[$ and which vanish at T. If $Y = \mathbb{R}$, we simply write $NBV(0,T)$ for $NBV(0,T;\mathbb{R})$. Let μ_v be the regular Borel measure belonging to $v \in NBV(0,T)$ and define on $[0,T] \times \bar{\Omega}$ a product measure by

$$\nu_v := \mu_v \times (dx),$$

where dx is the usual Lebesgue-measure on $\bar{\Omega}$.

THEOREM 2. Let the directional derivatives

$$f'(t,x,z;+1),\quad f'(t,x,z;-1)$$

of f with respect to the third variable z be bounded from above on every set $[0,T] \times \bar{\Omega} \times [a,b]$ with $a, b \in \mathbb{R}$. Let T be the optimal time and let u be an optimal control for (P). Suppose that there exists a control $\bar{u} \in L^\infty(\Sigma_T)$ with $|\bar{u}| \leq m$ and

$$\int_\Omega f(t,x,y(\bar{u})(t,x))\,dx < 0 \qquad \forall\, t \in [0,T].$$

Then there are $v \in NBV(0,T)$, a function q defined on $\bar{Q}_T$ which is summable with respect to ν_v and $h^* \in C(\bar{\Omega})^*$ with $h^* \neq 0$, $w^* \in L^1(Q_T)$, such that (2.1) and

$$(2.2)\quad \max_{k \in K} \langle k, h^*\rangle = \langle y(u)(T), h^*\rangle,$$

$$(2.3)\quad u\, w^*_{|\Sigma_T} + m|w^*_{|\Sigma_T}| = 0,$$

$$(2.4)\quad \int_0^T \int_\Omega f(t,x,y(u)(t,x))\,dx\,dv(t) = 0, \text{ } v \text{ is increasing,}$$

$$(2.5)\quad q(t,x) \in \partial f(t,x,y(u)(t,x)) \qquad \nu_v\text{ - a.e. on } \bar{Q}_T$$

hold. The functional z^* appearing in (2.1) is given by

$$(2.6)\quad \langle z,z^*\rangle = \int_0^T \int_\Omega z(t,x)q(t,x)\,dx\,dv(t) \qquad \forall\, z \in C(\bar{Q}_T).$$

Proof. First of all we show that the operator R defined by

$$(Rz)(t) \;.= \int_\Omega f(t,x,z(t,x))\,dx \qquad \forall\, t \in [0,T],\ \forall\, z \in C(\bar{Q}_T)$$

is continuous from $C(\bar{Q}_T)$ into $C(\bar{\Omega})$. For any $(t,x) \in \bar{Q}_T$ and any $y,\ z \in \mathbb{R}$ let

$$\omega(t,x,y,z) \;.= \max\,\{f'(t,x,y;\operatorname{sign}(y-z)),\ f'(t,x,z;\operatorname{sign}(z-y))\}.$$

For every $t \geq 0$, $x \in \bar{\Omega}$, $y,\ z \in \mathbb{R}$ with $y \neq z$ and any $h > 0$ we have

$$\frac{f(t,x,z) - f(t,x,z-h\,\operatorname{sign}(y-z))}{h} \leq \frac{f(t,x,y) - f(t,x,z)}{|y-z|}$$
$$\leq \frac{f(t,x,y+h\,\operatorname{sign}(y-z)) - f(t,x,y)}{h}$$

(compare Stoer/Witzgall [7], (4.1.4) Monotonicity Relations). Hence we conclude from the definition of the directional derivative

$$(2.7)\quad |f(t,x,y) - f(t,x,z)| \leq \omega(t,x,y,z)\,|y-z|.$$

Assume now that $\{z_i\}_{i\in\mathbb{N}}$ converges to z in $C(\bar{Q}_T)$. From our assumption and (2.7) we obtain

$$|f(t,x,z_i(t,x)) - f(t,x,z(t,x))| \leq c\|z_i - z\|_{C(\bar{Q}_T)}$$

with a constant $c > 0$. Hence $\{Rz_i\}_{i\in\mathbb{N}}$ converges to Rz in C[0,T].

We may now apply Theorem 2 of Mackenroth [4] and in this way we easily deduce the existence of $z^* \in C(\bar{Q}_T)^*$, $h^* \in C(\bar{\Omega})^* \setminus \{0\}$, $v \in NBV(0,T)$ and $w^* \in L^1(Q_T)$ with (2.1) - (2.4) and

$$z^* \in \partial r_v(y(u)),$$

where we have set

$$r_v(z) \;.= \int_0^T (Rz)(t)\,dv(t) \qquad \forall\, z \in C(\bar{Q}_T).$$

We calculate the conjugate r_v^* of the convex functional r_v. Clearly we can write

$$r_v(z) = \int_{\bar{Q}_T} f(s,z(s))\,d\nu_v \qquad \forall\, z \in C(\bar{Q}_T).$$

Assume that $\zeta \in C(\bar{Q}_T)^*$ can be represented in the form

$$(2.8)\quad \langle z,\zeta\rangle = \int_{\bar{Q}_T} z(s)p(s)\,d\nu_v \qquad \forall\, z \in C(\bar{Q}_T),$$

where p is a function which is summable on $\bar{Q}_T$ with respect to ν_v. From Rockafellar [5], Corollary 2, we obtain

$$r_v^*(\zeta) = \int_{\bar{Q}_T} f^*(s,p(s))\, d\nu_v.$$

If ζ is not of the form (2.8) it is also possible to apply this Corollary and one gets $r_v^*(\zeta) = +\infty$. The desired result follows now by calculating ∂r_v from r_v^* in the usual way. □

Remark 1. If f is differentiable with respect to z, the multiplier q is uniquely determined by v, namely be the equation

$$q(t,x) = f_z(t,x,y(u)(t,x)) \qquad \nu_v \text{ - a.e. on } \bar{Q}_T.$$

Remark 2. An analogous Theorem holds of course, if K is a subset of $L^2(\Omega)$ with a nonempty interior. Then we have $h^* \in L^2(\Omega)$.

III. BANG-BANG-CONTROLS

In this section we use Theorem 2 to prove a bang-bang-principle.

DEFINITION. Let $N \subset \Sigma_T$ be a set of positive measure. A control $u \in L^\infty(\Sigma_T)$ is said to be bang-bang on N, if $|u(t,\xi)| = m$ holds almost everywhere on N.

Essentially, we shall show that each optimal control for (P) is bang-bang on $M_u \times \Gamma$, where M_u is the set of all $t \in [0,T]$ such that the state constraint (1.4) is not active:

$$M_u := \{t \in [0,T] \mid \int_\Omega f(t,x,y(u)(t,x))\, dx < 0\}.$$

This is exactly the result that one expects from physical reasons. At first we have to prove a regularity result for the solution w* of the adjoint equation (2.1). To this end we need some definitions.

Let $C_o[0,T]$ be the subspace of all functions $\psi \in C[0,T]$ with $\psi(0) = \psi(T) = 0$ and put

$$\mathcal{A} := L^2(0,T;H^1(\Omega)^*) + \mathcal{L}(C_o[0,T], L^2(\Omega)).$$

$\mathcal{A}$ may be considered as a subspace of the space $\mathcal{D}^*(]0,T[;H^1(\Omega)^*)$ of distributions on $]0,T[$ with values in $H^1(\Omega)^*$. Let

$$\mathcal{B} := \{y \in L^2(0,T;H^1(\Omega)) \mid \frac{dy}{dt} \in \mathcal{A}\}.$$

By defining a norm in a natural way, $\mathcal{A}$ and $\mathcal{B}$ become Banach spaces (see Mackenroth [2]). Each $w^* \in \mathcal{B}$ is continuous on the right on $]0,T[$, continuous at 0 and T and of bounded variation, if it is considered as a function from $[0,T]$ into $H^1(\Omega)^*$. For any $t \in [0,T]$, $w^*(t)$ lies in $L^2(\Omega)$ and the map $w^* \mapsto w^*(t)$ is continuous from $\mathcal{B}$ into $L^2(\Omega)$ (cf. Mackenroth [2], Satz 4.2 and Satz 4.1).

Let $(\cdot,\cdot)$ be the inner product of $L^2(\Omega)$ and the duality of $H^1(\Omega)$ and $H^1(\Omega)^*$ (which is compatible). Any $z^* \in C([0,T];L^2(\Omega))^*$ may uniquely be written in the form

$$(3.1)\quad \langle z,z^*\rangle = \langle z,z_o^*\rangle + (z(0),z_o) + (z(T),z_T) \quad \forall\, z \in C([0,T];L^2(\Omega))$$

with $z_o, z_T \in L^2(\Omega)$ and where the function in $NBV(0,T;L^2(\Omega))$ representing $z_o^* \in C([0,T];L^2(\Omega))^*$ is continuous at 0 and T (cf. Mackenroth [2], Satz 5.1). This function is denoted by $v_{z_o^*}$. The functional z_o^* may be viewed as an element of $\mathcal{L}(C_o[0,T];L^2(\Omega))$ by defining

$$\rho(z_o^*)(\psi) := \int_0^T \psi \, dv_{z_o^*} \qquad \forall\, \psi \in C_o[0,T].$$

Finally we introduce an operator $\mathcal{O}\!\mathcal{C} \in \mathcal{L}(H^1(\Omega),H^1(\Omega)^*)$ by

$$a(v,w) := (w,\mathcal{O}\!\mathcal{C}v) \qquad \forall\, v,w \in H^1(\Omega),$$

where a is the bilinear form

$$a(v,w) := \sum_{i,j=1}^{n} (a_{ij}\frac{\partial v}{\partial x_i},\frac{\partial w}{\partial x_j}) + (a_o v,w) + \alpha\langle v_{|\Gamma}, w_{|\Gamma}\rangle_{L^2(\Gamma)} \qquad \forall\, v,w \in H^1(\Omega).$$

$\mathcal{O}\!\mathcal{C}$ is self-adjoint.

The desired regularity result for w^* makes use of the following assumption.

(A) The functional z^* given by (2.6) (which we also denote by $z^*_{q,v}$) can be extended by continuity to $C([0,T];L^2(\Omega))$ for all possible multipliers q,v.

In many cases it is possible to guarantee (A) by applying condition (2.5), e.g. if f is continuously differentiable with respect to z or if f is of the form

$$f(t,x,z) = b(t,x)\,|z - c(t,x)| - d(t,x) \quad \forall\, t \geq 0,\ \forall\, x \in \bar{\Omega},\ \forall\, z \in \mathbb{R}$$

with continuous functions b, c, d.

THEOREM 3. Assume that (A) and $h^* \in L^2(\Omega)$ hold. Then the solution w^* of (2.1) with z^* given by (2.6) lies in $\mathcal{B}$ and is also the solution of the equation of evolution

$$(3.2)\quad -\frac{dw^*}{dt} + \mathcal{O}\!\mathcal{C}w^* = \rho(z_o^*),$$

(3.3) $w^*(T) = z_T + h^*$

(where z^* is decomposed as in (3.1)).

Proof. From Mackenroth [2], Satz 5.2, we know that (3.2), (3.3) has a unique solution $\hat{w}^* \in \mathcal{B}$. We are ready if we can show $w^* = \hat{w}^*$. Let be given $\psi \in H^1(0,T)$, $v \in H^2(\Omega)$ with

$$\psi(0) = 0, \quad \alpha v_{|\Gamma} + \frac{\partial v}{\partial n_A} = 0.$$

Using Green's identity, we obtain

$$\begin{aligned}\int_0^T (\hat{w}^*(t), \mathcal{A}(\psi(t)v))\,dt &= \int_0^T a(\psi(t)v, \hat{w}^*(t))\,dt \\ &= \int_0^T (A(\psi(t)v), \hat{w}^*(t))\,dt \\ &\quad + \int_0^T \left(\alpha\psi(t)v_{|\Gamma} + \frac{\partial(\psi(t)v)}{\partial n_A}, \hat{w}^*(t)_{|\Gamma}\right)_{L^2(\Gamma)} dt \\ &= \int_0^T (A(\psi(t)v), \hat{w}^*(t))\,dt.\end{aligned}$$

Hence we get by Mackenroth [2], Satz 4.1, and by (3.2), (3.3)

$$\begin{aligned}\left\langle \frac{\partial(\psi v)}{\partial t} + A(\psi v), \hat{w}^* \right\rangle_{L^2(Q_T)} &= -\left(v, \frac{d\hat{w}^*}{dt}(\psi)\right) + (\psi(T)v, \hat{w}^*(T)) \\ &\quad + \int_0^T (\psi(t)v, \mathcal{A}\hat{w}^*(t))\,dt \\ &= \langle \psi v, z_o^* \rangle + (\psi(T)v, z_T + h^*) \\ &= \langle \psi v, z^* \rangle + (\psi(T)v, h^*).\end{aligned}$$

A density argument shows now that $\hat{w}^*$ also fulfils (2.1). Since the solution of (2.1) is unique, we must have $w^* = \hat{w}^*$. □

Assuming that (A) holds, (2.1) may be written in the form

$$\left\langle \frac{\partial\varphi}{\partial t} + A\varphi, w^* \right\rangle_{L^2(Q)} = \langle \varphi, z_o^* \rangle + \langle \varphi(T), z_T + h^* \rangle \qquad \forall\, \varphi \in B_T.$$

We denote the solution of this equation by $w^*(z_o^*, z_T + h^*)$. Since we have $w^*(z_o^*, 0) \in \mathcal{B}$ by Theorem 3 and $w^*(0, z_T + h^*) \in C([0,T[\times\bar{\Omega})$ by Mackenroth [3], Theorem 3, we are in a position to introduce the set

$$\Gamma(z^*) = \{h^* \in C(\bar{\Omega})^* \setminus \{0\} \mid \exists\, b \in\,]0,T[\, : w^*(z_o^*,0)(b) = w^*(0,h^*)(b)\}.$$

THEOREM 4. Suppose that the assumptions of Theorem 2 hold. Let u be an optimal control for (P) and let z^*, h^* be as in Theorem 2. Assume that assumption (A) is fulfilled. Then u is bang-bang on $M_u \times \Gamma$ if

(3.4) $z_T + h^* \notin \Gamma(-z^*) \cup \{0\}$.

Proof. Assume that u is not bang-bang on $M_u \times \Gamma$. Since M_u is an open set, it can be written as a countable union of intervals [a,b]. Hence there must be an interval $[a,b] \subset M_u$ with $b < T$ such that $|u(t,\xi)| < m$ for all (t,ξ) in a subset of $[a,b] \times \Gamma$ with positive measure. Since $w^*(z_o^*,0)$ lies in $\mathcal{B}$, we may assume without loss of generality that $w^*(z_o^*,0)$ is continuous at b. From Mackenroth [4], Theorem 3, we conclude

$$w^*(z_o^*,0)(t,x) = - w^*(0,z_T + h^*)(t,x) \qquad \text{a.e. on } [a,b] \times \Omega.$$

This implies

$$w^*(z_o^*,0)(b) = - w^*(0,z_T + h^*)(b),$$

which is a contradiction to (3.4). □

Because of $h^* \neq 0$, (3.4) always holds if (P) does not contain the state constraint (1.4). In the general situation, one may suspect that "normally" z^* and h^* are not linked in such a way that $w^*(z_o^*,0)(b)$ coincides with $-w^*(0,z_T + h^*)(b)$ for a $b \in]0,T[$ (in this case, h^* is uniquely determined by z^*, compare Theorem 5). Nevertheless, since z^* and h^* are not known in advance, one has to prove (3.4) for <u>all</u> possible multipliers z^*, h^*, if one wishes to guarantee that u is bang-bang on $M_u \times \Gamma$. In other words, one has to show that $z_T \neq - h^*$ and

(3.5) $\quad \Gamma(z^*_{q,v}) = \emptyset$ for all possible multipliers q, v with $v(T-) = 0$

hold. (3.5) is a little bit more than necessary, since h^* is restricted by (2.2).

An equivalent formulation of (3.5) may be given by using appropriate attainable sets. For $b \in]0,T[$ define

$$E(b) := \{w^*(z^*_{q,v},0)(b) \mid q,v \text{ are possible multipliers with } v(T-) = 0\},$$
$$E_T(b) := \{w^*(0,h^*)(b) \mid h^* \in C(\bar{\Omega})^* \setminus \{0\}\}.$$

Then (3.5) is equivalent to

(3.6) $\quad E(b) \cap E_T(b) = \emptyset \qquad \forall\, b \in]0,T[.$

The next Theorem shows that (3.6) (resp. (3.5)) is equivalent to the unsolvability of a certain moment problem. Denote by λ_k (resp. v_k) the eigenvalues (resp. eigenfunctions) of the eigenvalue problem

$$- A v_k + \lambda_k v_k = 0,$$
$$\alpha v_{k|\Gamma} + \frac{\partial v_k}{\partial n_A} = 0.$$

THEOREM 5. Condition (3.6) holds if and only if the moment problem

$$(3.7)\quad \int_b^T e^{\lambda_j(T-\tau)} \int_\Omega q(\tau,\xi)v_j(\xi)\,d\xi\,dv(\tau) = \langle v_j,h^*\rangle \quad \forall j \in \mathbb{N}$$

has no solution $h^* \in C(\bar{\Omega})^* \setminus \{0\}$ for every $b \in]0,T[$ and all possible multipliers q,v with $v(T-) = 0$.

Proof. From the proof of Mackenroth [3], Theorem 4, it may easily be deduced that we have for any function $\psi \in \mathcal{D}(]0,T[)$ and any $j \in \mathbb{N}$ (with $w^* := w^*(z^*_{q,v},-h^*)$)

$$\int_0^T \int_\Omega \psi(t)v_j(x)w^*(t,x)\,dx\,dt = \int_0^T \psi(t) \int_t^T e^{-\lambda_j(\tau-t)} \int_\Omega q(\tau,\xi)v_j(\xi)\,d\xi\,dv(\tau)\,dt - \int_0^T \psi(t)e^{-\lambda_j(T-t)}\,dt\,\langle v_j,h^*\rangle.$$

Hence we obtain for every $j \in \mathbb{N}$

$$\int_\Omega v_j(x)w^*(t,x)\,dx = \int_t^T e^{-\lambda_j(\tau-t)} \int_\Omega q(\tau,\xi)v_j(\xi)\,d\xi\,dv(\tau) - e^{-\lambda_j(T-t)}\langle v_j,h^*\rangle \text{ a.e. on } [0,T].$$

Since $w^* : [0,T] \to H^1(\Omega)$ is continuous on the right on $]0,T[$, we deduce from this equation that $w^*(b) = 0$ is equivalent to (3.7). Hence the Theorem is proved. □

In general it is difficult to decide whether the moment problem of Theorem 5 is solvable or not. We shall limit ourselves to the case where the state constraint is linear.

COROLLARY. Suppose that

$$f(t,x,z) = \rho(x)z \qquad \forall t \geq 0,\ \forall x \in \bar{\Omega},\ \forall z \in \mathbb{R}$$

with $\rho \in C(\bar{\Omega})$. Let K be a subset of $L^2(\Omega)$ having a nonempty interior. Let T be the optimal time and let u be an optimal control. Moreover, assume that there exists a control $\bar{u} \in L^\infty(\Sigma_T)$ such that the state constraint is strictly fulfilled for all $t \in [0,T]$. Then u is bang-bang on $M_u \times \Gamma$, if $z_T \neq -h^*$ for the multipliers z^*, h^*, and if

$$(3.8)\quad \sum_{j=1}^\infty (\rho,v_j)e^{2\lambda_j t} = +\infty \qquad \forall t \in]0,T[.$$

Proof. From Theorem 2 we have $q(\tau,\xi) = \rho(\xi)$ for all $(t,\xi) \in \bar{Q}_T$. We show that (3.8) implies that

(3.9) $$(\rho,v_j)\int_b^T e^{\lambda_j(T-\tau)}\,dv(\tau) = (v_j,h^*) \qquad \forall\, j \in \mathbb{N}$$

has no solution $h^* \in L^2(\Omega)\setminus\{0\}$ for every $b \in\,]0,T[$ and all increasing $v \in NBV(0,T)$ with $v(T-) = 0$. Then the assertion follows from Theorem 4 and Theorem 5. Let now such a b and v be given. Clearly we can assume $v_{|[b,T]} \neq 0$. Then there exists $t_o \in\,]b,T[$ with $v(b) < v(t_o)$. Denote the left side of (3.9) by α_j. Certainly the moment problem (3.9) has no solution $h^* \in L^2(\Omega)$ if

$$\sum_{j=1}^{\infty}\alpha_j^2 = +\infty.$$

We have the estimate

$$\begin{aligned}\int_b^T e^{\lambda_j(T-\tau)}\,dv(\tau) &\geq \int_b^{t_o} e^{\lambda_j(T-\tau)}\,dv(\tau)\\ &\geq e^{\lambda_j(T-t_o)}(v(t_o)-v(b)).\end{aligned}$$

Hence we obtain by (3.8)

$$\begin{aligned}\sum_{j=1}^{\infty}\alpha_j^2 &= \sum_{j=1}^{\infty}(\rho,v_j)^2\Big(\int_b^T e^{\lambda_j(T-\tau)}\,dv(\tau)\Big)^2\\ &\geq \sum_{j=1}^{\infty}(\rho,v_j)^2\, e^{2\lambda_j(T-t_o)}(v(t_o)-v(b))^2 = +\infty. \quad \square\end{aligned}$$

In some cases (3.8) can easily be verified, e.g. if we have $\rho = 1$, $n = 1$, $a_{11} = 1$ and $a_o = 0$, since in this situation the eigenfunctions v_j are explicitly known. In other cases, the moment problem (3.9) has a solution $h^* \in L^2(\Omega)$, e.g. if

$$\rho \in \text{span}\,\{v_1,\ldots,v_l\}$$

with $l \in \mathbb{N}$. Then by Theorem 5 condition (3.6) also fails.

In this context we have to mention Tröltzsch [8], where a bang-bang-principle for a similar problem with linear state constraints is derived. Using a different proof, he also obtains condition (3.8) (even a more general form, which applies to several linear constraints), but he does not look at sufficient conditions for u to be bang-bang on $M_u \times \Gamma$ of the type used in Theorem 4.

REFERENCES

1. J. L. Lions, Optimal Control of Systems Governed by Partial Differential Equations, Springer-Verlag, Berlin - Heidelberg - New York, 1971.

2. U. Mackenroth, Optimalitätsbedingungen und Dualität bei zustandsrestringierten parabolischen Kontrollproblemen, Math. Operationsforsch. Statist., Ser. Optimization 12 (1981), 65 - 89.

3. U. Mackenroth, Convex parabolic boundary control problems with pointwise state constraints, Journ. Math. Anal. Appl. 87 (1982), 256 - 277.

4. U. Mackenroth, Time-optimal parabolic boundary control problems with state constraints, Numer. Funct. Anal. and Optimiz. 3 (1981), 285 - 300.

5. R. T. Rockafellar, Integrals which are convex functionals, Pacific J. Math. 24 (1968), 525 - 539.

6. E. P. J. G. Schmidt, N. Weck, On the boundary behaviour of solutions to elliptic and parabolic equations - with applications to boundary control for parabolic equations, SIAM J. Control and Optimization 16 (1978), 593 - 598.

7. J. Stoer, C. Witzgall, Convexity and optimization in finite dimensions I, Springer-Verlag, Berlin - Heidelberg - New York, 1970.

8. F. Tröltzsch, On some parabolic boundary control problems with constraints on the control and functional-constraints on the state, to appear.

STATIC AND DYNAMICAL LOADS, POINTWISE CONSTRAINT IN STRUCTURAL OPTIMIZATION

B. Rousselet

Université de Nice
Nice, France

I. INTRODUCTION

The aim of the lecture is to derive necessary optimality conditions for weight minimization of structures under dynamic load (second order evolution equation); the minimization problem is subject to constraints on the thickness of the structure and a pointwise constraint on the response of the structure (heuristic design sensitivity as well as a numerical algorithm has already been given in Haug and Feng (1978)).

The same minimization problem is first considered under static load (elliptic equation); the result (already considered in Rousselet (1981b)) is interesting in itself and is also introductory to the dynamic case : the treatment of the pointwise constraint (which is generally nondifferentiable and non convex) in order to derive necessary optimality conditions is already intricate.

Eigenvalue constraint may lead also to nondifferentiable phenomena (see Haug and Choi (1981) and Rousselet (1981a & 1981b) but is not considered here.

Pionnering work in the area of distributed structural optimization is Armand (1972). A lot of introductory as well as advanced material is to be found in the Proceedings edited by Céa and Haug (1981). Lectures of Céa, Zolésio and the author provide a comprehensive self-contained treatment of shape optimal design (the design variable being the domain Ω itself occupied by the structure i.e. in which the partial differential equation is set), it is not considered here. Shape optimal design for dynamically loaded structures may be also considered (Rousselet (1982a)).

The need to consider singular behaviors in structural optimization is somewhat new but is expanding rapidly due to the practical interest of the pointwise constraints and repeated eigenvalues (see Olhoff and Rasmuzen (1977), Mazur (1980), Banichuk (1981)). On the other hand, a lot of papers have been published in abstract non differentiable optimization with few, if not any), connection to real differential equations ; see for example Aubin (1979) and Hiriart-Urruty (1978).

II. STATE EQUATIONS AND OPTIMIZATION PROBLEMS

II.1 State equations

To be concrete we consider the weight minimization of two specific examples of variable thickness structures: beam and plate. The bilinear forms are

- for the beam (of variable thickness, see fig.1)

$$a_u(z,v) = E\alpha \int_0^1 u^2 z_{xx} v_{xx}\, dx \tag{2.1}$$

where subscript x denotes derivative with respect to this variable, E is young's modulus (constant of the material), α is a positive constant (depending on the shape of the cross section (supposed to be all similar in shape); u is the variable cross-sectional area, presumed to be in $L^\infty(\Omega)$ (very smooth design are unnatural see Haug and Rousselet (1980a);

- for the plate (of variable thickness, see fig.2)
  (Ω is a bounded open subset of $\mathbb{R}^2$)

$$a_u(z,v) = \iint_\Omega D(u)\,[\, z_{11} v_{11} + z_{22} v_{22} + \nu\,(z_{11} v_{22} + z_{22} v_{11}) + 2(1-\nu)\, z_{12} v_{12}]\, d\Omega \tag{2.2}$$

where subscripts 1 (resp.2) denote derivatives with respect to x_1 (resp. x_2) and $D(u) = \dfrac{E u^3}{12(1-\nu^2)}$; E is still young's modulus, $\nu, |\nu| < \frac{1}{2})$ is Poisson's ratio and $u \in L^\infty(\Omega)$ denotes the (variable) thickness of the plate.

Remark. The space V which will be used to set the equation satisfied by the response of the structure is here the Sobolev space $H_0^2(\Omega)$; the choice of this space mechanically means that the structures are (so called) clamped ones; other boundary conditions could be considered as well; a_u is (in both cases) selfadjoint and V-coercive.

If A_u denotes the operator (from V to V') associated to the bilinear form a_u (using Riesz representation theorem), the displacement z of the structure under static load is solution of

$$A_u z = f \qquad (z \in V) \tag{2.3}$$

where $f \in L^2(\Omega)$ (or sometimes $f \in V'$ if one wants to include for example point loads (i.e. Dirac masses)) is the load acting on the structure. (2.3) is an abstract formulation of a fourth order differential equation with Dirichlet boundary condition.
The solution of (2.3) is indeed continuous (see e.g. theorem 1-9.8 of Lions-Magenes (1968)). Under dynamic load the displacement z of the structure is solution of

$$\rho u \frac{\partial^2 z}{\partial t^2} + A_u z = f \tag{2.4}$$

where $f \in L^2(0,T;\, L^2(\Omega))$ and $z \in L^2(0,T;\, V)$ (2.5)
and $\frac{\partial z}{\partial t} \in L^2(0,T;\, L^2(\Omega))$ with initial datas

$$z(0,\cdot) = z_o(\cdot),\ \frac{\partial z}{\partial t}(0,\cdot) = z_1(\cdot) \tag{2.6}$$
$$(z_o \in V,\, z_1 \in L^2(\Omega)).$$

This equation does not fit exactly in the classical frame (see theorem 3.8.1) in Lions-Magenes (1968)) because of the terms ρu in front of $\frac{\partial^2 z}{\partial t^2}$; however it may be proved that the same result holds (note that u does not depend on t); as expected from the mechanical origin of the problem it may be proved that, the solution is more regular in time (see theorem 3.8.2 in Lions-Magenes (1968) for the standard case): $z_u \in \mathscr{C}([0,T];V)$ and $\frac{\partial z_u}{\partial t} \in \mathscr{C}([0,T];L^2(\Omega))$. Moreover in our examples (Remark page 2), V is included in $H^2(\Omega)$, so that (see e.g. theorem 1.9.8 of Lions-Magenes (1968)), V is included in $\mathscr{C}(\bar{\Omega})$ and consequently $z_u \in \mathscr{C}([0,T] \times \bar{\Omega})$ (may be after modification on a set of measure zero in $\bar{\Omega}$).
Note that A_u actually involves derivatives of order four so that equation (2.4) is not an hyperbolic one but of Petrowsky type.

II.2 Minimization problems

The functional to be minimized is the weight of the structure :

$$J(u) = \rho \int_\Omega u(x)\, d\Omega \qquad \text{(linear functional)} \tag{2.7}$$

under various constraints : $0 \leqslant \alpha \leqslant u(x) \leqslant \beta$ a.e. in Ω (2.8)

and $K(z_u) \leqslant o$ (2.9)

where $K(z_u) = \sup_{x \in \bar{\Omega}} |z_u(x)|$ for static (2.10)

load and $K(z_u) = \sup_{(x,t) \in \bar{\Omega} \times [0,T]} |z_u(t,x)|$ for dynamic (2.11)

load , (the subscript u here emphasizes the dependence on u).

Constraint (2.8) is implicit with respect to u; to derive necessary optimality conditions or to implement (for example) a gradient projection type algorithm, one should be able to differentiate the mapping $u \longmapsto z_u$ (in other words to perform a design sensitivity analysis). Moreover (2.9) is pointwise (or local) as opposed to integral functionals such as the compliance of the structure

$$\frac{1}{2} \int_\Omega f\, z_u \, dx \ ; \tag{2.12}$$

if the supremum is reached at several points constraint (2.9) will be non differentiable.

III. DESIGN SENSITIVITY ANALYSIS

Design sensitivity analysis of static response has been rigorously proved in Haug-Rousselet (1980a) by proving the differentiability of A_u^{-1} ; this last result may be used to get design sensitivity analysis of repeated eigenvalues (Rousselet (1977) and Haug-Rousselet (1980b)) for systems which may not be selfadjoint.

III.1 Static loading

We provide here a proof of design sensitivity under static load which is simpler than the one to be found in Haug-Rousselet (1980a); it seems new but restricted to variational equations with coercivity assumptions on the bilinear form; it may be extended to dynamic load but (as a drawback) does not yield differentiability of A_u^{-1}.

Formal computation leads to define the derivative of z_u with respect to u denoted by $z'_{u,\delta u}$ as solution of :

$$\forall\, v \in V \quad a_u(z'_{u,\delta u}, v) = a'_{u,\delta u}(z_u, v) \tag{3.1}$$

where $a'_{u,\delta u}$ denotes the derivative of the bilinear form with respect to u :

- for a beam,

$$a'_{u,\delta u}(z,v) = 2E\alpha \int_0^1 u\, \delta u\, z_{xx}\, v_{xx}\, dx \ ; \tag{3.2}$$

- for a plate,

$$a'_{u,\delta u}(z,v) = \iint_\Omega D'(u,\bar{u}) [z_{11} v_{11} + z_{22} v_{22} + \nu (z_{11} v_{22} + z_{22} v_{11}) + 2(1-\nu)\, z_{12} v_{12}]\, d\Omega \tag{3.3}$$

where $D'(u,\bar{u}) = \frac{3Eu^2 \bar{u}}{12(1-\nu^2)}$ is the derivative of D(u) (with respect to u).

It is an exercise to show that $a'_{u,\delta u}$ is linear with respect to δu and that the following inequality holds for any $z \in V$, $v \in V$

$$|a'_{u,\delta u}(z,v)| \leq C \|\delta u\|_{L^\infty(\Omega)} \|z\|_V \, \|v\|_V \tag{3.4}$$

It implies that the right hand side of (3.1) is in V' which in turns yields existence and unicity of $z'_{u,\delta u} \in V$ (theorem 2.9.1 of Lions-Magenes (1968)). Moreover by slightly modifying the proof of theorem 3.1 of Haug-Rousselet (1980a) one can prove the following assertion :

PROPOSITION 3.1: Let a'_u be either defined in (2.1) or in (2.2), then the following estimate holds for any $z \in V$, any $v \in V$ and for $\|\delta u\|_{L^\infty(\Omega)}$ small enough :

$$|a_{u+\delta u}(z,v) - a_u(z,v) - a'_{u,\delta u}(z,v)| \leqslant \alpha(\delta u) \|z\|_V \|v\|_V \tag{3.5}$$

where $\alpha(\delta u) = o(\|\delta u\|_{L^\infty(\Omega)})$ (Landau notation)

(differentiability in the so called sense of relatively bounded perturbations).

THEOREM 3.2 : Let $z_u \in V$ be the solution of (2.3) (where A_u is associated either to (2.1) or to (2.2)), then z_u is differentiable with respect to u in supremum norm :

$$\sup_{x \in \bar{\Omega}} |z_{u+\delta u}(x) - z_u(x) - z'_{u,\delta u}| = o(\|\delta u\|). \tag{3.6}$$

To prove the estimate (3.6) one can first notice that due to the continuous imbedding of V into $\mathscr{C}(\bar{\Omega})$ (see e.g. theorem 1.9.8 of Lions-Magenes (1968)) it is sufficient to prove :

$$\|\delta^2 z_u\|_V = o(\|\delta u\|) \tag{3.7}$$

where to simplify the computations;

we set
$$\delta^2 z_u = z_{u+\delta u} - z_u - z'_{u,\delta u} \tag{3.8}$$

(2.3) in variational form is :

$$\forall\ v \in V \quad a_u(z_u,v) = (f,v) \tag{3.9}$$

Consequently :

$$\forall\ v \in V \quad a_{u+\delta u}(z_{u+\delta u},v) - a_u(z_u,v) = o \tag{3.10}$$

or with (3.8) :

$$a_{u+\delta u}(z_u + z'_{u,\delta u} + \delta^2 z_u, v) - a_u(z_u,v) = o\ . \tag{3.11}$$

Setting
$$\delta^2 a_u = a_{u+\delta u} - a_u - a'_{u,\delta u} \tag{3.12}$$

(3.11) may be expanded in the following way :

$$a_{u+\delta u}(\delta^2 z_u,v) + a_u(z_u + z'_{u,\delta u},v) + a'_{u,\delta u}(z_u + z'_{u,\delta u},v) + \delta^2 a_u(z_u + z'_{u,\delta u},v) - a_u(z_u,v) = o \tag{3.13}$$

expanding the second and third term and using the equation (3.1) of $z'_{u,\delta u}$

one gets :
$$a_{u+\delta u}(\delta^2 z_u,v) = -a'_{u,\delta u}(z'_{u,\delta u},v) - \delta^2 a_u(z_u + z'_{u,\delta u},v) \tag{3.14}$$

which should hold for any $v \in V$; consequently $\delta^2 z_u$ appears as the solution of an equation, the right hand side being (roughly speaking) of second order in δu, hence it should be also of second order in δu.

Precisely set $\varphi=\delta^2 z_u$ in (3.14), use uniform coercivity and continuity of the bilinear forms to get :

$$C_1 \|\delta^2 z_u\|_V^2 \leqslant C_2 \|\delta u\|_{L^\infty(\Omega)} \|z'_{u,\delta u}\|_V \|\delta^2 z_u\|_V + C_3 \|\delta u\|^2_{L^\infty(\Omega)} \|z_u + z'_{u,\delta u}\|_V \|\delta^2 z_u\|_V \tag{3.15}$$

Setting $v = z'_{u,\delta u}$ in (3.1) one gets easily

$$\|z'_{u,\delta u}\|_V \leqslant C_4 \|\delta u\|_{L^\infty(\Omega)} \tag{3.16}$$

Use (3.16) in (3.15) to get $\quad C_1 \|\delta^2 z_u\|_V \leqslant C_5 \|\delta u\|_{L^\infty(\Omega)}$ (3.17)

which implies the estimate (3.7) (even $\|\delta^2 z_u\|^2 = O(\|\delta u\|^2)$) and proves theorem 3.2.

III 2 <u>Dynamic load</u>

Rigorous proof of design sensitivity of variable thickness structures under dynamic load (second order evolution equation) seems new (see however Chavent (1971) for a related result under more restrictive smoothness assumptions on the coefficients of the equation).

The response of the structure is solution of (2.4) ; multiplying by $v\in V$, integrating over Ω and using a Green formula one gets :

$$\forall\ v\in V \quad \left(\rho u \frac{\partial^2 z_u}{\partial t^2}, v\right)_{V'\times V} + a_u(z_u,v)=(f,v)_H \tag{3.18}$$

with the initial datas (2.6)

As for the static case, a formal computation leads to define $z'_{u,\delta u}$ as the solution of :

$$\forall\ v\in V \quad (\rho u \frac{\partial^2}{\partial t^2} z'_{u,\delta u}, v) + a_u(z'_{u,\delta u}, v) = -\left(\delta u \frac{\partial^2 z_u}{\partial t^2}, v\right) - a'_{u,\delta u}(z_u,v) \tag{3.19}$$

with initials datas $z'_{u,\delta u}=0 \quad \frac{\partial}{\partial t} z'_{u,\delta u}=0.$ (3.20)

Note that the right hand side of (3.19) is not in $L^2(0,T;L^2(\Omega))$; the standard result (e.g. theorem 4.1.1 of Lions (1968)) does not apply; one could use solutions of a transposed equation (e.g. proposition 3.9.2 of Lions-Magenes (1968)); in that case $z'_{u,\delta u}$ would be in $L^2(0,T;L^2(\Omega))$ and not in $L^2(0,T;V)$ (this is typical of <u>second</u> order evolution equation) and some technical details would prevent the estimation of $\delta^2 z_u$ in $\mathscr{C}([0,T]\times\bar{\Omega})$.

Consequently the following additional assumptions (may be too strong) are considered : $\quad f\in\mathscr{C}^1([0,T];L^2(\Omega))$, $z_0\in D(A_u)$; $z_1\in V$ (3.21)
defined in (2.6))

then it may be proved by differentiating (2.4) with respect to time that

$$\frac{\partial^2 z}{\partial t^2}\in \mathscr{C}([0,T];L^2(\Omega)) . \tag{3.22}$$

These additional smoothness properties enable to show that $z'_{u,\delta u}$ considered as the solution of a transposition of equation (3.19) is indeed in $\mathscr{C}([0,T];V)$ and that $\|z'_{u,\delta u}\|_{\mathscr{C}([0,T];V)} \leqslant C\|\delta u\|_{L^\infty(\Omega)}$ (3.23)

As for the static case, the result of differentiability of z_u with respect to u which is needed to deal with the constraints of the optimization problem (see section 2.2) is :

$$\|\delta^2 z_u\|_{\mathscr{C}([0,T])\times\bar{\Omega})} = o(\|\delta u\|_{L^\infty(\Omega)}) \tag{3.24}$$

where
$$\delta^2 z_u = z_{u+\delta u} - z_u - z'_{u,\delta u}. \tag{3.25}$$

THEOREM 3.3 : Let z_u be the solution of (2.4) where A_u is the operator associated to the bilinear form a_u defined in (2.1) or in (2.2) ; assume additional smoothness assumptions (3.21) are satisfied then z_u is differentiable with respect to u (i.e. (3.24) is satisfied).

To prove theorem 3.3 eq.(3.18) is to be used for the values u and $u+\delta u$ of the design (but the same initial datas); substract both equations to get :

$$\left(\rho\,(u+\delta u)\,\frac{\partial^2 z_{u+\delta u}}{\partial t^2} - \rho\, u\,\frac{\partial^2 z_u}{\partial t^2}\,,v\right) = a_u(z_u,v) - a_{u+\delta u}(z_{u+\delta u},v) \tag{3.26}$$

As in the proof of theorem 3.2, eq.(3.26) is to be expanded using notation (3.25) and (3.12). Using equation (3.19), one gets :

$$\left(\rho\,(u+\delta u)\frac{\partial^2}{\partial t^2}\,\delta^2 z_u,v\right) + a_{u+\delta u}(\delta^2 z_u,v) = -\left(\rho\delta u\,\frac{\partial^2 z'_{u,\delta u}}{\partial t^2}\,,v\right) - a'_{u,\delta u}(z'_{u,\delta u},v)$$
$$- \delta^2 a_u(z_u + z'_{u,\delta u}\,,v) \tag{3.27}$$

Roughly speaking, the right hand side of eq.(3.27) being of order two in δu, this will imply that $\delta^2 z_u$ is also of order two. But as for equation (3.19), the right hand side of the last equation is not in $L^2(0,T\,;L^2(\Omega))$ so that one should consider at first glance $\delta^2 z_u$ as the solution of a transposition of equation (3.27). In fact one can prove directly that $\delta^2 z_u \in \mathscr{C}([0,T];V)$ (which is also a consequence of (3.25)) and that

$$\|\delta^2 z_u\|_{\mathscr{C}([0,T];V)} = o(\|\delta u\|_{L^\infty(\Omega)})$$

which is the expected estimate and proves the design sensitivity of the dynamic response.

IV. NECESSARY OPTIMALITY CONDITIONS

IV.1 Static case

The minimization problem to be considered is (2.7),(2.8),(2.9) and (2.10) with state equation (2.3) where A_u is the operator associated to the bilinear form of equation (2.1) or (2.2).

Pointwise constraint (2.1) is non differentiable if the supremum is reached at several points. To derive necessary optimality conditions at some local optimum u_*, the dual of the tangent cone of constraint (2.7) is to be computed (generalizing the gradient of a differentiable constraint).

This method is based on the ideas of Girsanov (1972) and Ioffe-Tihomirov (1979); it is closely connected to the peridifferential of Clarke, (see Lemaréchal (1980b)).

This dual cone is made of $g' \in (L^\infty(\Omega))'$ satisfying :

$$\langle g', \bar{u} \rangle = \int_\Omega z'_{u,\bar{u}}(x)\, d\mu(x) \tag{4.1}$$

for any $\bar{u} \in L^\infty(\Omega)$, where μ is some (regular, Borel) measure on Ω; moreover, if μ^+ and μ^- denote its positive and negative parts $(\mu = \mu^+ - \mu^-)$ (4.2)

it satisfies $\int_\Omega d\mu^+ + \int_\Omega d\mu^- = 1,$ (4.3)

where μ^+ is concentrated (in the sense of measure theory) on the set

$$S_+(z_u) = \{\xi \in \Omega \mid z_u(\xi) = \operatorname{Sup}\{|z_u(x)| \mid x \in \bar{\Omega}\}\} \tag{4.4}$$

and μ^- on the set

$$S_-(z_u) = \{\xi \in \Omega \mid z_u(\xi) = -\operatorname{Sup}\{|z_u(x)| \mid x \in \bar{\Omega}\}\} \tag{4.5}$$

The proof is technical; it is sketched in Rousselet (1981b); a detailed proof is to be found in Rousselet (1982a). In these papers one finds the following necessary condition :

THEOREM 4.1 : Consider the optimization problem (2.5),(2.6),(2.7) and (2.8) with state equation (2.3) (A_u associated to the bilinear form of formula (2.1) or (2.2)); let u_* be some local optimum, then there exists some positive numbers λ_0, λ_1 (not both zero) and a regular, Borel) measure μ which satisfies (4.3),(4.4) and (4.5) such that the following inequality holds

$$\lambda_0 \int_\Omega (u - u_*)\, dx + \lambda_1 \int_\Omega z'_{u_*, u-u_*}\, d\mu(x) \geq 0 \tag{4.6}$$

for any $u \in L^\infty(\Omega)$ which satisfies

which satisfies $0 < \alpha \leq u(x) \leq \beta$ a.e. in Ω (4.7)

A more explicit necessary condition may be obtained by introducing an *adjoint state* (as for smooth integral constraints. The adjoint state $p_{u_*;\mu}$ is here solution of $a_{u_*}(\varphi; p_{u_*;\mu}) = \int_\Omega \varphi(x)\, d\mu(x)$ (4.8) which should hold for any $\varphi \in V(\Omega)$ (1). Then selecting $\varphi = z'_{u_*,u-u_*}$ in (4.8)

$$\int_\Omega z'_{u_*,u-u_*} \, d\mu(x) = a_{u_*}(z'_{u_*,u-u_*}, p_{u_*;\mu}) \tag{4.9}$$

and using equation (3.1) (of definition of $z'_{u,\delta u}$) one gets :

$$\int_\Omega z'_{u_*;u-u_*} \, d\mu(x) = - \, a'_{u_*,u-u_*}(z_{u_*}, p_{u_*;\mu}) \tag{4.10}$$

and consequently :

COROLLARY 4.2 : Under the hypothesis of theorem 4.1, the necessary condition (4.6) may be replaced by

$$\lambda_o \int_\Omega (u-u_*)dx - \lambda_1 \, a'_{u_*,u-u_*}(z_{u_*}, p_{u_*;\mu}) \geqslant o \tag{4.11}$$

where $p_{u_*;\mu}$ is solution of (4.8).

Remark. Based on this necessary condition, an algorithm has been described in Rousselet(1981a). The use of general algorithms designed for non differentiable optimization in $\mathbb{R}^n$ (such as in so called « MODULOPT » club, see Lemarechal (1980a)) is under study; direct and adjoint state are then solved using finite elements; equation (4.11) may be used as a stopping criteria.

IV.2 Dynamic case

IV.2.1 *Tutorial example.*

As a model problem, the optimization problem (2.7),(2.8) with constraint :

$$\int_{\Omega \times [o,T]} g(z_u) \, dx \, dt \leqslant c \tag{4.12}$$

is considered where g is some smooth function (eq. (4.12) is a differentiable constraint) and z_u is solution of (2.4)(2.5) with initial datas (2.6) and with additional smoothness assumptions (3.21).

The usual Lagrange multiplier rule in its classical way (Courant-Hilbert (1953), Ioffe & Tihimirov (1979)) yields :

PROPOSITION 4.3 : Consider the optimization problem (2.7),(2.8),(4.12) with state equation (2.4),(2.5),(2.6) (where A_u is associated to the bilinear form a_u defined in (2.1) or (2.2)); let u_* be some local optimum, then there exists some positive numbers λ_o, λ_1 (not both zero) such that :

$$\lambda_o \int_{\Omega \times [o,T]} (u-u_*) \, dx \, dt + \lambda_1 \int_{\Omega \times [o,T]} g'(z_{u_*}) z'_{u_*,u-u_*} \, dx \, dt \geqslant o \tag{4.13}$$

for any $u \in L^\infty(\Omega)$ which satisfies (4.7).

(1) The subscript $u_*;\mu$ used for p just emphasises the dependence of p on u_* and μ as opposed to the subscript u, δu used for z' which means the derivative of z computed at u in the direction δu.

For g smooth $g'(z_{u_*}) \in L^2(o,T\,;L^2(\Omega))$ (even $L^2(o,T;V)$) and the following adjoint equation fits in the classical frame (theorem 3.8.1 in Lions-Magenes (1968)) slightly modified (see the discussion of eq.(2.4) in section 2.1) :

$$\rho u \frac{\partial^2 p_{u*}}{\partial t^2} + A_{u_*} p_{u_*} = g'(z_{u_*}) \tag{4.14}$$

$$p_{u_*}(T) = o = \frac{\partial p_{u_*}}{\partial t}(T) \tag{4.15}$$

where $p_{u_*} \in L^2(o,T;V)$, $\frac{\partial p_{u*}}{\partial t} \in L^2(o,T;L^2(\Omega))$ (4.16)

As for the static case, multiply eq.(4.14) by ψ integrate parts in time and space variables to get (using initial datas and boundary conditions):

$$\int_o^T \int_\Omega \rho u_* \, p_{u_*} \frac{\partial^2 \psi}{\partial t^2} \, dx\, dt + \int_o^T a_{u_*}(p_{u_*},\psi) dt = \int_o^T \int_\Omega g'(z_{u_*}) \psi \, dx\, dt \tag{4.17}$$

set $\psi = z'_{u_*,u-u_*}$ and use a transposition of eq.(3.19) (for $z'_{u,\delta u}$) to get :

$$\int_o^T \int_\Omega g'(z_{u_*}) \, z'_{u_*,u-u_*} dx\, dt = -\int_o^T a'_{u_*,u-u_*}(z_{u_*},p_{u_*}) dt - \int_o^T \int_\Omega \rho\,(u-u_*) \frac{\partial^2 z_{u*}}{\partial t^2} \, p_{u*} \, dx\, dt \tag{4.18}$$

Equation (4.18) yields an easier formulation of proposition 4.3 :

COROLLARY 4.4 : Under the hypothesis of proposition 4.3, necessary condition (4.13) may be replaced by :

$$\lambda_o \int_o^T \int_\Omega (u-u_*) dx\, dt - \lambda_1 \int_o^T a'_{u_*,u-u_*}(z_{u_*},p_{u_*}) dt - \lambda_1 \int_o^T \int_\Omega \rho (u-u_*) \frac{\partial^2 z_{u*}}{\partial t^2} p_{u_*} \, dx\, dt \geqslant o \tag{4.19}$$

where p_{u_*} is solution of (4.14),(4.15),(4.16) .

IV.2.2 *Pointwise constraints (for dynamic load)*

The optimization problem to be considered is now (2.7),(2.8),(2.9),(2.11) with state equation (2.4),(2.5), (2.6), (3.21) .

Constraint (2.9),(2.11) is a pointwise one, therefore it is non differentiable (if the supremum is reached at several points).

The schedule of the derivation is similar to the case of an integral constraint (as explained in the tutorial example) but the details of the proof are much more technical; some technical details have been considered when dealing with static response but some others are different due to the nature of the evolution equation governing the response of the structure.

Let u_* be a local optimum of the optimization problem considered at the begining of the present subsection.

The dual of the tangent cone of constraint (2.9),(2.11) is similar to the cone of constraint (2.9),(2.10) of the static case (see section 4.1). It is made up of $g' \in (L^\infty(\Omega))'$ satisfying

$$\langle g',\bar{u}\rangle = \int_o^T \int_\Omega z'_{u,\bar{u}}(x,t) d\mu(x,t) \tag{4.20}$$

for any $\bar{u} \in L^\infty(\Omega)$, where μ is some (regular, Borel) measure on $[0,T]\times\bar{\Omega}$; if μ^+ and μ^- denote its positive and negative parts ($\mu = \mu^+ - \mu^-$), it satisfies (compare with section 4.1) : $\int_o^T \int_\Omega d\mu^+ + \int_o^T \int_\Omega d\mu^- = 1$ (4.21)

μ^+ is concentrated on the set :

$$S_+(z_u) = \{(\xi,\tau) \in \bar{\Omega} \times [0,T] \mid z_u(\xi,\tau) = -\operatorname{Sup}\{|z_u(x,t)| \,|\, (x,t) \in \bar{\Omega}\times[0,T]\}\} \tag{4.22}$$

and μ^- on the set :

$$S_-(z_u) = \{(\xi,\tau) \in \bar{\Omega} \times [0,T] \mid z_u(\xi,\tau) = -\operatorname{Sup}\{|z_u(x,t)| \,|\, (x,t) \in \bar{\Omega}\times[0,T]\}\} \tag{4.23}$$

The following theorem may be considered as an extension of proposition 4.3 to the case of pointwise constraints as well as an extension of theorem 4.1 to dynamic response :

THEOREM 4.5 : Consider the optimization problem (2.7),(2.8),(2.9),(2.11) with state equation defined by (2.4),(2.5),(2.6) with smoothness hypothesis (3.21) (operator A_u associated to the bilinear form of formula (2.1) or (2.2)); let u_* be some local optimum, then there exists some positive numbers λ_o,λ_1 (not both zero) and a (regular, Borel) measure μ which satisfies (4.21),(4.22),(4.23) such that the following inequality holds :

$$\lambda_o \int_o^T \int_\Omega (u-u_*)\,dx\,dt + \lambda_1 \int_o^T \int_\Omega z'_{u_*,u-u_*}(x,t)\,d\mu(x,t) \geqslant o \tag{4.24}$$

for any $u \in L^\infty(\Omega)$ which satisfies $o < \alpha \leqslant u(x) \leqslant \beta$ a.e. in Ω.

A more explicit necessary condition may be obtained by introducing an adjoint state

$$\begin{cases} \rho u \dfrac{\partial^2 p_{u_*;\mu}}{\partial t^2} + A_{u_*} p_{u_*;\mu} = \mu & (4.25) \\ p_{u_*;\mu}(T) = o = \dfrac{\partial p_{u_*;\mu}(T)}{\partial t} & (4.26) \end{cases}$$

Obviously this equation should be interpreted by *transposition* (see theorem 3.9.2 of Lions-Magenes (1968)); the right hand side μ is actually in Ξ' (dual of Ξ defined in formula (3.9.9) of Lions-Magenes (1968)); $P_{u_*;u} \in L^2(0,T;L^2(\Omega))$.

As for the tutorial example (subsection 4.2.1), multiply equation (4.25) by ψ, integrate by parts in time and space variables to get (using initial datas and boundary conditions):

$$\int_o^T \int_\Omega \rho u_* p_{u^*;\mu} \frac{\partial^2 \psi}{\partial t^2}\,dx\,dt + \int_o^T a_{u_*}(p_{u_*;\mu},\psi)\,dt = \int_o^T \int_\Omega z'_{u_*,u-u_*}\,d\mu(x,t) \tag{4.27}$$

set $\psi = z'_{u_*,u-u_*}$ (this is possible because due to additional smoothness assumption (3.21), $z'_{u_*,u-u_*}$ belongs to $\mathscr{C}([0,T];V)$ and may be considered as a test function in the previous equation) and use a transposition of eq.(3.19) (for $z'_{u,\delta u}$) to get :

$$\int_0^T\int_\Omega z'_{u_*,u-u_*}\,d\mu(x,t) = -\int_0^T a'_{u_*,u-u_*}(z_{u_*},\, p_{u_*;\mu})dt - \int_0^T\int_\Omega \rho\,(u-u_*)\frac{\partial^2 z_{u_*}}{\partial t^2}\, p_{u_*;\mu}\,dx\,dt \tag{4.28}$$

equation (4.28) yields an easier formulation of theorem 4.5 (eq.(4.29) is only "formal" at this point; to give it a precise meaning some additional smoothness properties is needed).

COROLLARY 4.5 : Under the hypothesis of theorem 4.5, necessary condition (4.24) may be replaced by :

$$\lambda_o\int_0^T\int_\Omega(u-u_*)dx\,dt - \lambda_1\int_0^T a'_{u_*,u-u_*}(z_{u_*},p_{u_*;\mu})dt - \lambda_1\int_0^T\int_\Omega\rho(u-u_*)\,\frac{\partial^2 z_{u_*}}{\partial t^2}\, p_{u_*;\mu}\,dx\,dt \geqslant o \tag{4.29}$$

where $p_{u_*;\mu}$ is solution of (4.25),(4.26).

Remark of the end of section 4.1 applies as well to the dynamic case.

V. CONCLUSION

The lecture has provided derivation of design sensitivity formulas for dynamically loaded structures (second order evolution equations) which is an extension of a new simple derivation of design sensitivity of statically loaded structures (elliptic equation).

This precise design sensitivity enables to derive necessary optimality conditions for weight minimization of structures with pointwise constraints (in supremum norm) on the response for statically well as dynamically loaded structures.

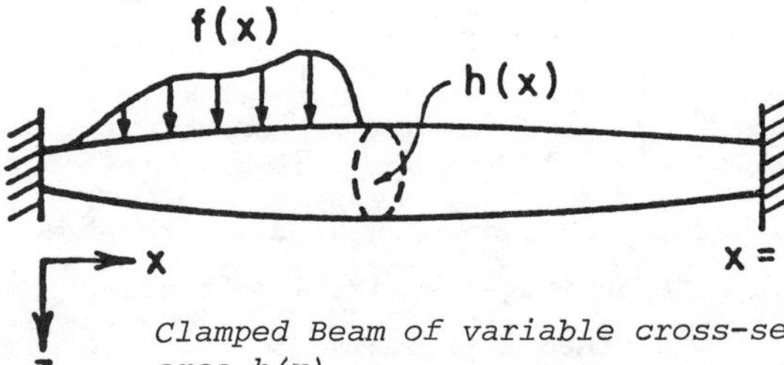

Figure 1 *Clamped Beam of variable cross-sectional area h(x)*

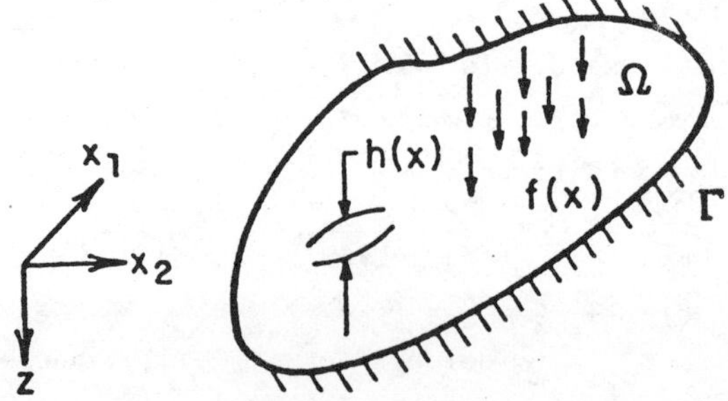

Figure 2 *Clamped Plate of variable thickness h(x)*

REFERENCES

Armand, J.L., (1972), Applications of the theory of optimal control of distributed-parameter systems to structural optimization, NASA Contractor Report, n° NASA-2044, NASA, Washington, D.C.

Aubin, J.P., (1979), Cônes tangents à un sous-ensemble convexe fermé, Ann. Sci. Math., Québec, 3, 63-80

Banichuk, N., (1981), Optimality conditions and analytical methods of shape optimization, Lecture in Céa, J. & E.J.Haug, ed., (1981)

Céa, J. & E.J. Haug ed. (1981), Proceedings of Nato Advanced Study Inst. on Optimization of Distributed Parameter Structural Systems, Iowa City, Iowa, U.S.A., 1980; to appear 1981 Sythoff & Noordhoff, Alphen aan den Rijn, The Netherlands

Chavent, G., (1971), Analyse fonctionnelle et identification de coefficients répartis dans les équations aux dérivées partielles, Thèse d'Etat, Paris

Courant, R., & D. **Hilbert**, (1953), Methods of Mathematical Physics, Interscience publishers

Girsanov, I.V., (1972), Lectures on Mathematical Theory of Extremum Problems, Springer Verlag, Berlin

Haug, E.J., & K. **Choi**, (1981), Optimization of structures with repeated eigenvalues, Lecture in Céa J., and E.J. Haug, ed. (1981)

Haug, E.J., & T. **Feng**, (1978), Optimal design of dynamically loaded continuous structures, Int. J. of Num. Meth. in Eng., 12, 299-307

Haug, E.J., & J.S. **Arora**, (1979), Applied Optimal Design, Wiley, New York

Haug, E.J. & B. **Rousselet** (1980a), Design sensitivity analysis in structural Mechanics I : Static Response Variations, J. of Struct. Mech., 8, 17-41

Haug, E.J., & B. **Rousselet**, (1980b), Design sensitivity analysis in structural Mechanics II, Eigenvalue variations, J. of Struct. Mech., 8, 161-186

Hiriart-Urruty, J.B., (1978), Gradients généralisés de fonctions marginales, SIAM J. Contr. Opt., 16, 301-316

Ioffe, A.D. & V.M. **Tihomirov**, (1979), Theory of extremal problems, North Holland, Amsterdam

Lemaréchal, C.,(1980a), Normes Modulopt I, Rapport INRIA, Rocquencourt, France

Lemaréchal, C.,(1980b), Nondifferentiable optimization, Lect.8 in Nonlinear Optimization, Theory and Algorithms, Dixon, Spedicato, Szegö eds. Birkhaüser, Boston

Lions, J.L., (1968), Contrôle optimal de systèmes gouvernés par des équations aux dérivées partielles, Dunod, Gauthier-Villars, Paris

Lions, J.L. & E. **Magenes**, (1968), Problèmes aux limites nonhomogènes et applications, vol.1 et 2, Dunod, Paris

Mazur, E.F., (1980), On singular optimality conditions in structural design, Proceedings, IUTAM Symp. on Variational Methods in the Mechanics of Solids, to appear

Olhoff, N., & S.H. **Rasmussen**, (1977), On single and bimodal buckling modes of clamped columns, Int. J. of Solids and Structures, 13, 605-614

Rousselet, B., (1977), Identification de domaines de problèmes de valeurs propres, Thèse, 3° cycle, Univ. de Nice, France

Rousselet, B. (1981a), Multiple eigenvalues (numerical algorithm) and supremum norm constraints, Lecture in Céa, J. & E.J. Haug, eds., (1981)

Rousselet, B., (1981b), Structural optimization; derivation of necessary conditions in some singular cases, Proceedings of the 2nd Workshop on control applications of nonlinear Programming and Optimization, Oberaffenhofen, RFA, (1980), to appear

Rousselet, B., (1982a), Conditions nécessaires d'optimalité : situations singulières en optimisation de structures, to appear

Rousselet, B., (1982b), Dynamic response in shape optimization, Proceedings of the 3rd IFAC Symp. on the Control of distributed parameter System, (June 29th-July 2nd,1982, Toulouse, France), Pergamon Press, (to appear)

ESTIMATION AND CONTROL IN FINITE STATE DISCOUNTED DYNAMIC PROGRAMMING

Manfred Schäl
University of Bonn
Bonn, West Germany

ABSTRACT

The present paper is concerned with Markov decision models with finite state space, arbitrary action space where the law of motion is not completely known. The principle of simultaneous estimation and (adaptive) control studied independently by KURANO and MANDL for the average return criterion is used here for the discounted return criterion. Bounds are established for the expected difference between the conditionally expected total discounted return from some stage outwards and the optimal value function. Also the Bayes approach is included. Moreover, the rate of convergence of the maximum likelihood estimation procedure and of the a posteriori distributions in Markov decision models is shown to be exponentially fast.

I. THE MODEL

The underlying decision model is a Markov decision model augmented by auxiliary observations which may yield information on the unknown parameter θ. Thus, estimators for θ may depend on the auxiliary observations. As the optimal policy will again be based on estimators for θ, policies depending on the auxiliary observations emerge.

We will write $\mathbb{R}$ for the set of real numbers, $\mathbb{N}_0$ and $\mathbb{N}$ for the set of nonnegative and positive integers, respectively, 1_B for the indicator function of a set B, and $E[f,B]$ for the expectation of f over B, i.e. $E[f,B] = E(f \cdot 1_B)$. For the finite set Θ, $\mathbb{P}(\Theta)$ is the set of all probability measures on Θ. The underlying *decision model* is given by a tupel $(I, Z, A, D, \Theta, q, r, \beta)$ of the following meaning:

1.1 I stands for the *state space* and is assumed to be finite.

1.2 Z is the *space of auxiliary observations* and is assumed to be denumerable.

1.3 A is the *action space*, which is an arbitrary space.

1.4 D associates with any $i \in I$ a nonempty subset of A, the *set of admissible actions* $D(i)$ if the system is in state i. We set $K = \{(i,a), i \in I, a \in D(i)\}$.

1.5 Θ is the *parameter space* and is assumed to be finite.

1.6 The so-called *transition law* q is a transition probability from $\Theta \times K$ to $Z \times I$. There $q^\theta(z,j|i,a)$ is the probability that the next auxiliary observation is z and the next state of the system is j evaluated for the parameter θ given that the system is in state i and the action a is taken. Then q defines the transition probabilities $p_{ij}^\theta(a) = \sum_{z \in Z} q^\theta(z,j|i,a)$, where $p_{ij}^\theta(a)$ is the probability that the next state of system is j given θ, i, and a as above.

1.7 For the *reward function* $r: \Theta \times K \to \mathbb{R}$ we assume that r is bounded.

1.8 The *discount factor* β is a number in the open unit interval.

We write $H_0 = I$, $H_{n+1} = K \times Z \times H_n$, $n \in \mathbb{N}_0$, $H_\infty = \times_0^\infty (K \times Z)$ with generic element $h = (i_0, a_0, z_1, i_1, a_1, \ldots)$. Let ξ_n, $n \in \mathbb{N}_0$, denote the projection from H_∞ on the n-th I-coordinate, ζ_n, $n \in \mathbb{N}$, the projection on the n-th Z-coordinate, α_n, $n \in \mathbb{N}_0$, the projection on the n-th A-coordinate and η_n the projection on H_n. Then the random variables ξ_n, ζ_n, α_n, η_n describe the state of the system at stage n, the auxiliary observation in period n, the action chosen at stage n and the history experienced up to stage n, respectively.

We agree to look on a function on H_n also as a function on H_∞ depending on $(i_0, a_0, z_1, i_1, a_1, \ldots) \in H_\infty$ only through $(i_0, a_0, \ldots, z_n, i_n)$.

A policy $\delta = (\delta_n, n \in \mathbb{N}_0)$ is defined as a sequence of functions δ_n from H_n to A such that $\delta_n(i_0, a_0, \ldots, z_n, i_n) \in D(i_n)$. We write Δ for the set of all policies. Given some initial distribution p on I, the transition law q^θ evaluated for some $\theta \in \textcircled{N}$ and a policy δ define a probability measure P_δ^θ on H_∞ and thus a random process $(\xi_o, \alpha_o, \zeta_1, \xi_1, \alpha_1, \ldots)$ (cp. Hinderer (1970) p. 80, Schäl (1981)). In this paper we are concerned only with the conditional distributions $P_\delta^\theta(\cdot|\eta_t)$ which are given by q^θ and δ and are independent of p. Therefore we can dispense with an initial distribution p. We will write $P_{i\delta}^\theta$ for $P_\delta^\theta(\ldots|\eta_o = i)$.

We call f a <u>decision function</u> and write $f \in \mathbb{F}$ iff f is a function $f: I \to A$ such that $f(i) \in D(i)$ for $i \in I$. We call δ a <u>Markov policy</u> iff $\delta = (\delta_n) = (f_n)$ where $f_n \in \mathbb{F}$, such that $f_n \circ \xi_n = \delta_n \circ \eta_n$, $n \in \mathbb{N}_o$. We identify $\mathbb{F}$ with the set of all <u>stationary policies</u>, i.e., $\delta \in \mathbb{F}$ or $\delta = f$ for some $f \in \mathbb{F}$ iff $\delta = (\delta_n)$ is chosen such that $f \circ \xi_n = \delta_n \circ \eta_n$. Under $f \in \mathbb{F}$, $\theta \in \textcircled{N}$, $(\xi_n, n \in \mathbb{N}_o)$ forms a homogeneous Markov chain with transition probabilities $(p_{ij}^\theta(f),\ i, j \in I)$, where $p_{ij}^\theta(f) = p_{ij}^\theta(f(i))$.

The optimality criterion of this paper will be based on the total discounted reward. Therefore we define

$$R_T^\theta(\delta) = E_\delta^\theta\left[\sum_{t=T}^\infty \beta^{t-T} r^\theta(\xi_t, \alpha_t)|\eta_T\right]$$

as the expected total discounted reward in $[T,\infty)$ discounted on time T given the history η_T. The <u>optimal value function</u> u^θ is defined by

$$u^\theta(i) = \sup_{\delta \in \Delta} R_o^\theta(\delta)(i),\ i \in I.$$

A policy δ^* is called <u>discount optimal</u> for θ iff $R_o^\theta(\delta^*) = u^\theta$.

1.9 PROPOSITION. (Principle of optimality, cp. Hinderer (1970), Theorem 3.8) $\sup_{\delta \in \Delta} R_T^\theta(\delta) = u^\theta \circ \xi_T$.

δ^* is discount optimal for θ iff

$$R_T^\theta(\delta^*) = u^\theta \circ \xi_T \quad P_{i\delta^*}^\theta \quad \text{- a.s. for all } T \in \mathbb{N},\ i \in I.$$

We will be concerned with the case where θ is not known to the decision maker. Therefore, one may ask whether there is a policy independent of θ which is good for <u>all</u> θ. Indeed, this has been shown for the so-called average reward criterion (cp. Kurano 1972, Mandl 1974, Kolonko 1980,1981a). The reason behind it is that the average cost depends only on the asymptotic behaviour of the process. But in the far future, on has observed the process for a long time and may have good information on the underlying 'true' parameter. Then the choice of the actions may be based on this information which will result in good actions.

The situation is completely different for the total discounted reward which heavily depends on the actions in the first periods. But then one has little information on the true parameter and the chosen actions may be very bad. Thus, for the discounted reward criterion one cannot expect that there exists a policy which is optimal for all θ. For that reason we modify our optimality criterion such that on the one hand it is based on the total discounted reward and on the other hand it depends only on the asymptotic behaviour of the process. Indeed, we will look for conditions that the identities in 1.9 approximately hold for large T. Our main goal is to obtain bounds for

$$E^{\theta}_{i\delta}|R^{\theta}_{T}(\delta) - u^{\theta} \circ \xi_T| \tag{1.10}$$

for finite T which tend to zero as T tends to infinity.

II. ESTIMATION

2.1 DEFINITION. $(\hat{\theta}_m, m \in \mathbb{N}_o)$ is an <u>asymptotic estimator</u> for θ iff $\hat{\theta}_m$ is a function from H_m into Ⓝ, $m \in \mathbb{N}_o$.

The problem of finding good estimators in a sequential decision model was studied by Kurano (1972), Mandl (1974), Borkar and Varaiya (1979,1980), Doshi and Shreve (1980), Kolonko (1980, 1981b) and Abdel-Fattah (1980). In more special examples, e.g. in the control of queues, good estimators are given in a natural way (cp. Kolonko (1980, 1981a), Kolonko, Schäl (1980), Schäl (1981)).

Besides estimators $\hat{\theta}_n$, one can also use the a posteriori distributions $\hat{\mu}_n : H_n \to \mathbb{P}$(Ⓝ) defined for a given a priori distribution $\hat{\mu}_o \in \mathbb{P}$(Ⓝ) to describe the status of information given at stage n after having experienced the history η_n.

2.2 DEFINITION. a) Let T : $\mathbb{P}$(Ⓝ) × K × Z × I → $\mathbb{P}$(Ⓝ) be defined by $T(\theta|\mu, i, a, z, j) = c^{-1} \cdot q^{\theta}(z, j|i, a)\mu(\theta)$ if $c = \sum_{\theta'} q^{\theta'}(z, j|i, a)\mu(\theta') > 0$ and $T(\theta|\mu, i, a, z, j) = \mu(\theta)$ if $c = 0$.

b) The sequence of a posteriori distributions $(\hat{\mu}_n)$ is defined for a given a priori distribution $\hat{\mu}_o \in \mathbb{P}(\textcircled{N})$ recursively by $\hat{\mu}_{n+1}(h_n, a, z, j) = T(\hat{\mu}_n(h_n), i_n, a, z, j)$ for $h_n = (\ldots, i_n) \in H_n$.

2.3 LEMMA. $\sum_{\theta'} \mu(\theta') q^{\theta'}(z, i|i, a) T(\theta|\mu, i, a, z, j) = \mu(\theta) q^{\theta}(z, j|i, a)$ if c as defined 2.2 is positive.

The proof is obvious.

We introduce some operators and quantities. For any u: I → ℝ set

$$L^{\theta}u(i, a) = r^{\theta}(i, a) + \beta \sum_{j \in I} p^{\theta}_{ij}(a) u(j), \quad a \in D(i)$$

$$\phi^{\theta}(i, a) = L^{\theta}u^{\theta}(i, a) - u^{\theta}(i), \quad U^{\theta}u(i) = \max_{a \in D(i)} L^{\theta}u(i, a), \quad i \in I.$$

Set $\overline{b} = \sup_{(i, a) \in K, \theta \in \textcircled{N}} r^{\theta}(i, a)/(1-\beta)$, $\underline{b} = \inf\ldots$ $b = \overline{b} - \underline{b}$.

Let us summarize some well-known facts from dynamic programming (cp. Bertsekas, Shreve (1978) I, 4, Schäl (1981)).

2.4 PROPOSITION. (a) $\underline{b} \le R^{\theta}_T(\delta) \le \overline{b}$, $\underline{b} \le u^{\theta} \le \overline{b}$, $\underline{b} \le L^{\theta}u^{\theta} \le \overline{b}$, $-b \le \phi^{\theta} \le 0$

(b) Optimality equation: $u^{\theta} = U^{\theta}u^{\theta}$, i.e., $\sup_{a \in D(i)} \phi^{\theta}(i, a) = 0$;

(c) Optimality criterion: f is discount optimal for θ iff

$$\phi^{\theta}(i, f(i)) = 0 \quad \text{for all } i \in I.$$

(d) $$E^{\theta}_{i\delta}|R^{\theta}_T(\delta) - u^{\theta} \circ \xi_T| = \sum_{t=T}^{\infty} \beta^{t-T} E^{\theta}_{i\delta}|\phi^{\theta}(\xi_t, \alpha_t)|$$

We learn from 2.4c that $\phi^{\theta}(i, a)$ is a measure of goodness of the action a (if the system is in state i and θ is the true parameter) and from 2.4d that the quantity we are mainly interested in can be measured in terms of ϕ^{θ}. When the information on θ is based on the a posteriori distribution $\mu \in \mathbb{P}(\textcircled{N})$, say, we have to look for a substitute for $\phi^{\theta}(i, a)$ which is defined for (μ, i, a). At first we consider the so-called Bayes equivalence principle introduced by van Hee (1978) and define

$$\phi^{\mu}(i, a) = \sum_{\theta} \mu(\theta) \phi^{\theta}(i, a) - \sup_{a' \in D(i)} \sum_{\theta} \mu(\theta) \phi^{\theta}(i, a')$$

As ϕ^{θ}, the quantity ϕ^{μ} enjoys the properties

(2.5) $$\phi^{\mu}(i, a) \le 0, \quad \sup_{a \in D(i)} \phi^{\mu}(i, a) = 0.$$

Finally, we will study the Baysian approach and define

$$u^*(\mu, i) = \sup_{\delta \in \Delta} \sum_{\theta} \mu(\theta) R_o(\delta, i), \qquad i \in I, \ \mu \in \mathbb{P}(\textcircled{N}).$$

Then u* is the optimal value function for the Bayesian decision model and is the unique bounded solution of the Bayesian optimality equation (cp. Rieder (1975), van Hee (1978)).

(2.6) $u^*(\mu, i) = \sup_{a \in D(i)} Lu^*(\mu, i, a)$ where

$$Lu^*(\mu, i, a) = \sum_\theta \mu(\theta)\{r^\theta(i, a) + \beta \sum_{j,z} q^\theta(z, j|i, a)u^*(T(\mu,i,a,z,j),j)\}$$

Now we define

$$\phi(\mu, i, a) = Lu^*(\mu, i, a) - u^*(\mu, i).$$

Then, again we have

(2.7) $\phi(\mu, i, a) \leq 0; \sup_{a \in D(i)} \phi(\mu, i, a) = 0.$

The following bounds for u^* will be useful (cp. van Hee (1978, section 5):

2.8 LEMMA. $\mu(\theta_o)u^{\theta_o} + (1 - \mu(\theta_o))\underline{b} \leq u^*(\mu, i) \leq \sum_\theta \mu(\theta)u(\theta).$

The proof immediately follows from the definition of u^*. It is clear from the theory of Bayesian Markov decision models that according to the same reasoning as above one may look on $\phi(\mu, i, a)$ as a measure of goodness of the action a if the system is in state i and μ is the a posteriori distribution. The quantity $\phi^\mu(i, a)$ is a substitute for $\phi(\mu, i, a)$ where $\phi^\mu(i, a)$ is much easier to calculate than $\phi(\mu, i, a)$.

The next lemma establishes lower bounds for the goodness of some action a in state i with respect to the true parameter θ_o in terms of the goodness w.r.t. an estimated value θ or w.r.t. the a posteriori distribution μ, say.

2.9 LEMMA. Let be $(i, a) \in K$, $\theta, \theta_o \in Ⓝ$, $\mu \in \mathbb{P}(Ⓝ)$:

(a) $\phi^{\theta_o}(i, a) \geq \phi^\theta(i, a) - b\, 1_{[\theta \neq \theta_o]}$,

(b) $\phi^{\theta_o}(i, a) \geq \phi^\mu(i, a) - 2b(1 - \mu(\theta_o))$

(c) $\phi^{\theta_o}(i, a) \geq \phi(\mu, i, a) - 2b(1 - \mu(\theta_o)).$

Proof. Part a) is clear from 2.4a.
b) From the definition of ϕ^μ and 2.4a,c we obtain

$$\phi^{\mu}(i, a) \le \mu(\theta_o)\phi^{\theta_o}(i, a) - \max_{a'}\{\mu(\theta_o)\phi^{\theta_o}(i, a')$$
$$- b(1 - \mu(\theta_o))\} = \mu(\theta_o)\phi^{\theta_o}(i, a) + b(1 - \mu(\theta_o))$$
$$\le \phi^{\theta_o}(i, a) + 2b(1 - \mu(\theta_o)).$$

c) By 2.8, 2.3 it follows that

$$Lu^*(\mu, i, a) \le \sum_\theta \mu(\theta)\{r^\theta(i, a)$$
$$+ \beta\sum_{j,z} q^\theta(z, j|i, a)\sum_{\theta'} T(\theta'|\mu, i, a, z, j)u^{\theta'}(j)\}$$
$$= \sum_\theta \mu(\theta)r^\theta(i, a)$$
$$+ \beta\sum_{\theta'}\mu(\theta')\sum_{j,z} q^{\theta'}(z, j|i, a)u^{\theta'}(j)$$
$$= \sum_\theta \mu(\theta)L^\theta u^\theta(i, a).$$

Hence, by 2.8, 2.4a,

$$\phi(\mu, i, a) \le \mu(\theta_o)L^{\theta_o}u^{\theta_o}(i, a) + (1 - \mu(\theta_o))\overline{b}$$
$$- \mu(\theta_o)u^{\theta_o}(i) - (1 - \mu(\theta_o))\underline{b}$$
$$= \mu(\theta_o)\phi^{\theta_o}(i, a) + b(1 - \mu(\theta_o))$$
$$\le \phi^{\theta_o}(i, a) + 2b(1 - \mu(\theta_o)). \qquad \square$$

2.10 THEOREM. Let $(\hat{\theta}_\mu)$ be an asymptotic estimatior for θ and let $(\hat{\mu}_n)$ be the sequence of a posteriori distributions defined for a given a priori distribution $\hat{\mu}_o$. Then each of the following expressions is an upper bound for

$$E^\theta_{i\delta}|R^\theta_T(\delta) - u^\theta \circ \xi_T| \quad \text{where } \delta \in \Delta,\ i \in I,\ T \in \mathbb{N}_o,\ \theta \in \text{Ⓝ}:$$

1. $$\frac{1}{1-\beta}\sup_{t\ge T}\{E^\theta_{i\delta}|\phi^{\hat{\theta}}t(\xi_t, \alpha_t)| + bP^\theta_{i\delta}(\hat{\theta}_t \ne \theta)\}$$

2. $$\frac{1}{1-\beta}\sup_{t\ge T}\{E^\theta_{i\delta}|\phi^{\hat{\mu}}t(\xi_t, \alpha_t)| + 2b[1 - E^\theta_{i\delta}\hat{\mu}_t(\theta)]\}$$

3. $$\frac{1}{1-\beta}\sup_{t\ge T}\{E^\theta_{i\delta}|\phi(\hat{\mu}_t, \xi_t, \alpha_t)| + 2b[1 - E^\theta_{i\delta}\hat{\mu}_t(\theta)]\}$$

The proof immediately follows from 2.4d and 2.9. Kolonko (1982) obtained bounds for case 1 in 2.10 also if the parameter space is some metric space under Lipschitz-type continuity assumptions.

The upper bounds established in 2.10 can be written as the sum of two error terms. The first one is small if the actions are good with respect to the estimated values for θ or w.r.t. the a posteriori distributions. In view of 2.4b, 2.5 and 2.7, this first error term can be made arbitrarily small by the choice of δ.

For example, choose $\varepsilon_t > 0$, $t \in \mathbb{N}_0$, such that $\varepsilon_t \to 0$. Further, choose $\delta_t(h_t)$, $t \in \mathbb{N}_0$, such that $\phi^\theta(i_t, \delta_t(h_t)) \geq -\varepsilon_t$ for $\theta = \hat{\theta}_t(h_t)$ and all $h_t = (\ldots, i_t) \in H_t$. Then

$$\sup_{t \geq T} E^\theta_{i\delta} |\phi^{\hat{\theta}} t(\xi_t, \alpha_t)| \leq \sup_{t \geq T} \varepsilon_t \to 0 \ (T \to \infty).$$

The second error term will become small if $(\hat{\theta}_t)$ is uniformly consistent:

2.11 DEFINITION. An asymptotic estimatior $(\hat{\theta}_m)$ is called <u>uniformly consistent</u> iff $(\hat{\theta}_m)$ converges to θ in probability with respect to $P^\theta_{i\delta}$ i.e., $P^\theta_{i\delta}(\hat{\theta}_m \neq \theta) \to 0$ as $m \to \infty$ for all $i \in I$, $\delta \in \Delta$, and $\theta \in Ⓝ$.

2.12 DEFINITION. δ^* is <u>asymptotically discount optimal</u> for θ iff $E^\theta_{i\delta^*}|R^\theta_T(\delta^*) - u^\theta \circ \xi_T| \to 0$ as $T \to \infty$ for all $i \in I$.

2.13 COROLLARY. If there exist a uniformly consistent asymptotic estimator for θ, then there exists a policy which is asymptotically discount optimal for all $\theta \in Ⓝ$.

It is remarkable that the Corollary holds without compactness of $D(i)$, $i \in I$.

In section 4 we will give upper bounds for the second error terms in 2.10 and sufficient conditions for the uniform consistency of the maximum likelihood estimators and for the convergence of the a posteriori distributions which implies that $1 - E^\theta_{i\delta}\hat{\mu}_t(\theta) \to 0$.

In most papers on estimation and control (Schäl (1982) is an exception) the case is considered where δ is defined through a set $\{f^\theta, \theta \in Ⓝ\} \in \mathbb{F}^{Ⓝ}$ and an asymptotic estimator:

2.14 DEFINITION. For a given set of stationary policies $\{f^\theta, \theta \in Ⓝ\}$ and an asymptotic estimator $(\hat{\theta}_m)$, we say: δ is adapted to $(\hat{\theta}_m)$ and $\{f^\theta, \theta \in Ⓝ\}$ iff

$$\delta_t = f^{\hat{\theta}_t} \circ \xi_t, \qquad t \in \mathbb{N}_0$$

i.e. $$\delta_t(i_0, a_0, \ldots, z_t, i_t) = f^{\hat{\theta}_t(i_0, a_0, \ldots, z_t, i_t)}(i_t).$$

In the situation described in 2.14 one has

$$E_{i\delta}^{\theta}|\phi^{\hat{\theta}_t}(\xi_t, \alpha_t)| \le \sup_{i,\theta}|\phi^{\theta}(i, f^{\theta}(i))| \quad .$$

Since by 2.4c

$$\inf_{\{f^{\theta},\ \theta \in Ⓝ\} \in \mathbb{F}^{Ⓝ}} |\phi^{\theta}(i, f^{\theta}(i)| = 0 \quad ,$$

the last bound can be made arbitrarily small by a suitable choice of $\{f^{\theta},\ \theta \in Ⓝ\}$.

III. A LIMIT THEOREM

In some cases, not only the limit of the difference 1.10 but the limit of $E_{i\delta}^{\theta} R_T^{\theta}(\delta)$ as $T \to \infty$ can be evaluated.

3.1 LEMMA. Let be $u: I \to \mathbb{R}$, $(\hat{\theta}_m)$ an asymptotic estimator, $\{f^{\theta},\ \theta \in Ⓝ\} \in \mathbb{F}^{Ⓝ}$ and let δ be adapted to $(\hat{\theta}_m)$ and $\{f^{\theta},\ \theta \in Ⓝ\}$. Then for $n > M$:

$$E_{\delta}^{\theta}\left[u \circ \xi_n, \hat{\theta}_m = \theta, m > M | \eta_M\right] = E_{f^{\theta}}^{\theta}\left[u \circ \xi_n, \hat{\theta}_m = \theta, m > M | \eta_M\right] \quad \text{on } [\hat{\theta}_M = \theta] \quad .$$

Proof. It is sufficient to prove for $l \in \mathbb{N}$

$$E_{\delta}^{\theta}\left[u_l(\xi_{M+1},\ldots,\xi_{M+l}),\ \hat{\theta}_m = \hat{\theta},\ M < m \le M + l | \eta_M\right]$$
$$E_{f^{\theta}}^{\theta}\left[u_l(\zeta_{M+1},\ldots,\xi_{M+l}),\ \hat{\theta}_m = \theta,\ M < m \le M + l | \eta_M\right] \quad \text{on } [\hat{\theta}_M = \theta]$$

for any bounded function u_l. This can easily be carried through by induction on l. □

3.2 LEMMA. Let be $(\hat{\theta}_m)$, $\{f^{\theta},\ \theta \in Ⓝ\}$, δ as in 2.8. Then

$$\left| E_{i\delta}^{\theta}\left[R_N^{\theta}(\delta)\right] - E_{i\delta}^{\theta}\left[E_{f^{\theta}}^{\theta}[R_N^{\theta}(f^{\theta}) | \eta_M]\right]\right|$$

$$\le \quad b \cdot \kappa_i^{\theta}(M) \qquad \text{for } M < N \text{ where}$$

$$\kappa_i^{\theta}(M) = \sup_{\sigma \in \Delta} P_{i\sigma}^{\theta}\left(U_{m \ge M}[\theta_m \neq \theta]\right) \quad .$$

Proof.

$E_{i\delta}^{\theta}[R_N^{\theta}(\delta)]$

$$= \sum_{n=N}^{\infty}\beta^{n-N}\ E_{i\delta}^{\theta}\Big[r^{\theta}(\xi_n,\ f^{\theta}(\xi_n)),\ \hat{\theta}_m = \theta,\ m \geq M\Big] + r_{MN}^{1}$$

$$= \sum_{n=N}^{\infty}\beta^{n-N}\ E_{i\delta}^{\theta}\Big[E_{\delta}^{\theta}[r^{\theta}(\xi_n,\ f^{\theta}(\xi_n)),\ \hat{\theta}_m = \theta,\ m > M|\eta_M],\ \hat{\theta}_M = \theta\Big] + r_{MN}^{1}$$

$$= \sum_{n=N}^{\infty}\beta^{n-N}\ E_{i\delta}^{\theta}\Big[E_{f\theta}^{\theta}\ [r^{\theta}(\xi_n,\ f^{\theta}(\xi_n)),\ldots,m > M|\eta_M],\ \hat{\theta}_M = \theta\Big] + r_{MN}^{1}$$

$$= E_{i\delta}^{\theta}\Big[E_{f^{\theta}}^{\theta}[\sum_{n=N}^{\infty}\beta^{n-N}\ r^{\theta}(\xi_n,\ f^{\theta}(\xi_n))|\eta_M]\Big] + r_{MN}^{1} + r_{MN}^{2}$$

$$= E_{i\delta}^{\theta}\Big[E_{f^{\theta}}^{\theta}[R_N^{\theta}(f^{\theta})|\eta_M]\Big] + r_{MN}^{2} + r_{MN}^{1}$$

where use was made of 3.1 and r_{MN}^{1}, r_{MN}^{2} can be defined in an appropriate way such that

$$r_{MN}^{1} + r_{MN}^{2} \leq b \cdot \kappa_i^{\theta}(M),\ i = 1,\ 2. \qquad \square$$

3.3 LEMMA. Let be $f \in \mathbb{F}$, $\theta \in \textcircled{N}$, $(p_{ij}^{\theta}(f),\ i,\ j \in I)$ an irreducible aperiodic Markov transition matrix with stationary distribution π^{θ}. Then for all $i \in I$

$$\lim_{N\to\infty} E_{if}^{\theta} R_N^{\theta}(f) = (1-\beta)^{-1}\sum_{j\in I}\pi^{\theta}(j)r^{\theta}(j,\ f(j)).$$

Proof.

$$E_{if}^{\theta} R_N^{\theta}(f) = \sum_{n\geq N}\beta^{n-N}\sum_{j\in I} r^{\theta}(j, f(j))\, P_{if}^{\theta}(\xi_n = j) \qquad .$$

By assumption we have $\lim_{n\to\infty} P_{if}^{\theta}(\xi_n = j) = \pi^{\theta}(j)$. Now the assertion follows. $\square$

3.4 THEOREM. Suppose that $\{f^{\theta},\ \theta \in \textcircled{N}\}$ is given such that $(p_{ij}^{\theta}(f^{\theta}),\ i,\ j \in I)$ are irreducible aperiodic Markov transition matrices with stationary distributions π^{θ}. Further suppose that $(\hat{\theta}_m)$ is an asymptotic estimator such that

$$\kappa_i^{\theta}(M) = \sup_{\sigma\in\Delta} P_{i\sigma}^{\theta}\Big[\bar{\cup}_{m\geq M}[\hat{\theta}_m \neq \theta]\Big] \to 0 \text{ as } M \to \infty \qquad .$$

Let δ be adapted to $(\hat{\theta}_m)$ and $\{f^{\theta},\ \theta \in \textcircled{N}\}$. Then

$$\lim_{n\to\infty} E_{i\delta}^{\theta} R_N^{\theta}(\delta) = (1-\beta)^{-1}\sum_{j\in I}\pi^{\theta}(j)r^{\theta}(j,\ f^{\theta}(j)) \qquad .$$

Sufficient conditions for $\kappa_i^{\theta}(M) \to 0$ are given in section 4. Other situations where $\kappa_i^{\theta}(M) \to 0$ arise if the distribution of $(\zeta_n,\ n \in \mathbb{N})$ is independent of the underlying policy and $\hat{\theta}_m$ depends only on $(\zeta_1,\ldots,\zeta_m)$. Then

$P^{\theta}_{i\sigma}(U_{m\in M}[\theta_m \neq \theta])$ is independent of σ and converges to zero if $\hat{\theta}_m \to \theta$ P^{θ}_{i}-almost sure.

Proof of 3.4. By 3.2 it is sufficient to evaluate

$$\lim_{N\to\infty} E^{\theta}_{i\delta}\left[E^{\theta}_{f^{\theta}}[R_N(f^{\theta})\,|\,\eta_M]\right]$$

For the stationary policy f^{θ} we have

$$E^{\theta}_{f^{\theta}}\left[R^{\theta}_N(f^{\theta})\,|\,\eta_M\right] = E^{\theta}_{f^{\theta}}\left[R^{\theta}_{N-M}(f^{\theta})\cdot\eta_{N-M}\,|\,\eta_o\right]\circ\xi_M \to (1-\beta)^{-1}\sum_{j\in I}\pi^{\theta}(j)r^{\theta}(j,f(j))$$

by 3.3. Now the assertion follows from the dominated convergence Theorem.

IV. BOUNDS FOR THE ESTIMATION PROCEDURE

4.1 DEFINITION. Set $q^{\theta}_n(i_o, a_o,\ldots, z_n, i_n) = \prod_{1\le m\le n} q^{\theta}(z_m, i_m|i_{m-1}, a_{m-1})$. The asymptotic estimatior $(\hat{\theta}_n)$ is a maximum likelihood estimatior iff $\hat{\theta}_n(h_n)$ is a maximum point of the function $\theta \to q^{\theta}_n(h_n)$ for all $h_n \in H_n$, $n \in \mathbb{N}$. Throughout this section we make the following additional assumption.

4.2 ASSUMPTION. a) $q^{\theta}(z, j|i, a)$ is continuous in a on $D(i)$ for all i, $j\in I$, $z\in Z$, $\theta\in\Theta$.

b) A is a metric space and $D(i)$ is compact for all $i\in I$.

c) For θ, $\theta'\in\Theta$ where $\theta \neq \theta'$ there exist a set $J(\theta, \theta')\subset I$ such that

1. $q^{\theta}(\cdot|j, a) \neq q^{\theta'}(\cdot|j, a)$ for all $a\in D(j)$, $j\in J(\theta, \theta')$;
2. there exists $n = n(\theta, \theta')\in\mathbb{N}$ such that $P^{\theta''}_{i\delta}(\tau > n) < 1$ for $\theta''\in\{\theta, \theta'\}$, $i\in I$, $\delta\in\Delta$ where $\tau = \tau(\theta, \theta') = \inf\{m\in\mathbb{N}_o, \xi_m\in J(\theta, \theta')\}$.

4.3 DEFINITION. $\bar{n} = \max\{n(\theta, \theta'), \theta \neq \theta'\} + 1$

$d = \sup_{i,\delta,\theta\neq\theta'}\max\{P^{\theta''}_{i\delta}(\tau(\theta, \theta') \ge \bar{n}), \theta''\in\{\theta, \theta'\}\}$.

4.4 LEMMA. $d < 1$

Proof. Δ is compact with respect to the topology of pointwise convergence (cp. Derman (1965), Schäl (1975)) and it is easily verified that $\delta \to P^{\theta}_{i\delta}(\tau > n)$ is continuous. □

4.5 DEFINITION. For $\theta \neq \theta'$, $\rho_{ia}(\theta, \theta')$ is the affinity

$$\sum_{z,k}\{q^{\theta}(z, j|i, a) \cdot q^{\theta'}(z, j|i, a)\}^{1/2}$$

and

$$\rho = \sup_{\theta \neq \theta', j \in J(\theta, \theta'), a \in D(j)} \rho_{ja}(\theta, \theta'). \ \lambda = \rho + (1-\rho)d.$$

4.6 LEMMA. $\rho < 1$.

Proof. We have $\rho_{ia}(\theta, \theta') < 1$ by Schwarz's inequality. Since ρ_{ia} is continuous in a, the result follows. □

4.7 DEFINITION. $Q_n(\theta, \theta') = \{q^{\theta'}(\zeta_n, \xi_n|\xi_{n-1}, \alpha_{n-1})/q^{\theta}(\zeta_n, \xi_n|\xi_{n-1}, \alpha_{n-1})\}^{1/2}$.
$X_n(\theta, \theta') = \prod_{1 \leq m \leq n} Q_m(\theta, \theta')$, $n \in \mathbb{N}$, $X_0 = 1$.

4.8 LEMMA. (a) $E_{\delta}^{\theta}[Q_{n+1}(\theta, \theta')|\eta_n] \leq \rho$ on $[\xi_n \in H(\theta, \theta')]$ and ≤ 1 elsewhere, $\theta \neq \theta'$, $\delta \in \Delta$, $n \in \mathbb{N}$,

(b) $X_n(\theta, \theta')$ is a positive supermartingale under $P_{i\delta}^{\theta}$ for all $i \in I$, $\delta \in \Delta$, $\theta \neq \theta'$,

Proof. It is easy to see that

$$E_{\delta}^{\theta}[Q_{n+1}|\theta, \theta')|\eta_n] = \rho_{\xi_n, \delta_n(\eta_n)}(\theta, \theta').$$

Now (a) follows and (b) is a consequence of (a). □

4.9 LEMMA. $E_{i\delta}^{\theta} X_{\bar{n}}(\theta, \theta') \leq \lambda < 1$ for all i, δ, $\theta \neq \theta'$

Proof. Set $\tau = \tau(\theta, \theta')$, $X_n = X_n(\theta, \theta')$, $\tau \wedge n = \min(\tau, n)$. Then it follows from 4.8b and then 4.8a that

$$\begin{aligned} E_{i\delta}^{\theta} X_{\bar{n}} &= \sum_{m=0}^{\bar{n}-1} E_{i\delta}^{\theta}[X_{\bar{n}}, \tau = m] + E_{i\delta}^{\theta}[X_{\bar{n}}, \tau \geq \bar{n}] \\ &\leq \sum_{m=0}^{\bar{n}-1} E_{i\delta}^{\theta}[X_{m+1}, \tau = m] + E_{i\delta}^{\theta}[X_{\bar{n}}, \tau \geq \bar{n}] \\ &\leq \rho \sum_{m=0}^{\bar{n}-1} E_{i\delta}^{\theta}[X_m, \tau = m] + E_{i\delta}^{\theta}[X_{\bar{n}}, \tau \geq \bar{n}] \\ &= \rho E_{i\delta}^{\theta}[X_{\tau}, \tau < \bar{n}] + E_{i\delta}^{\theta}[X_{\bar{n}}, \tau \geq \bar{n}] \end{aligned}$$

$$= \rho E_{i\delta}^{\theta}[X_{\tau \wedge \bar{n}}] + (1 - \rho)E_{i\delta}^{\theta}[X_{\bar{n}}, \tau \geq \bar{n}]$$

$$\leq \rho + (1 - \rho)E_{i\delta}^{\theta}[X_{\bar{n}-1}, \tau \geq \bar{n}]$$

Finally, we obtain from Schwarz's inequality

$$E_{i\delta}^{\theta}[X_{\bar{n}-1}, \tau \geq \bar{n}] \leq \{P_{i\delta}^{\theta}(\tau \geq \bar{n}) \cdot P_{i\delta}^{\theta'}(\tau \geq \bar{n})\}^{1/2}$$

$$\leq d. \qquad \square$$

4.10 LEMMA. $E_{i\delta}^{\theta}X_n(\theta, \theta') \leq \lambda^{-1} \cdot \lambda^{n/\bar{n}}$ for all $i, \delta, \theta \neq \theta', n \in \mathbb{N}$.

Proof. Write $n = m\bar{n} + l$, $0 \leq l \leq \bar{n}$ and extend 4.9 by induction on m. $\square$

4.11 LEMMA. $P_{i\delta}^{\theta}(U_{\theta' \neq \theta}[q_n^{\theta'} \circ \eta_n \geq \alpha^n \cdot q_n^{\theta} \circ \eta_n])$

$$\leq (|\mathbb{N}| - 1) \cdot \lambda^{-1} \cdot (\lambda^{1/\bar{n}}/\alpha^{1/2})^n,$$

for all $i, \delta, \theta \neq \theta', n \in \mathbb{N}, \alpha > 0$.

Proof. Using the Markov inequality, one obtains

$$\sum_{\theta' \neq \theta} P_{i\delta}^{\theta}(q_n^{\theta} \circ \eta_n \geq \alpha^n q_n^{\theta} \circ \eta_n)$$

$$= \sum_{\theta' \neq \theta} P_{i\delta}^{\theta}(X_n(\theta, \theta') \geq \alpha^{n/2})$$

$$\leq \sum_{\theta' \neq \theta} \alpha^{-n/2} E_{i\delta}^{\theta} X_n(\theta, \theta'). \qquad \square$$

4.12 THEOREM. If $(\hat{\theta}_n)$ is a maximum likelihood estimator, then

$$P_{i\delta}^{\theta}(\hat{\theta}_n \neq \theta) \leq (|\mathbb{N}| - 1) \cdot \lambda^{-1} \cdot \lambda^{n/\bar{n}}$$

$$P_{i\delta}^{\theta}(U_{m \geq n}[\hat{\theta}_n \neq \theta])) \leq (|\mathbb{N}| - 1)(\lambda(1 - \lambda^{1/\bar{n}}))^{-1}\lambda^{n/\bar{n}}$$

for all i, δ, θ, n.

Proof. Use 4.11 for $\alpha = 1$. $\square$

The method of proof of 4.12 was already used by Rényi (1964).

Theorem 4.12 generalizes a result by Kurano (1972) where $H(\theta, \theta') = I$ for all $\theta \neq \theta'$. Mandl (1974) considers the case where $\textcircled{N}$ is any compact set and $H(\theta, \theta')$ is a single ton for all $\theta \neq \theta'$.

4.13 THEOREM. Let be $(\hat{\mu}_n)$ the sequence of a posteriori disributions defined for a given a priori distribution $\hat{\mu}_o$ and set $\alpha = \lambda^{2/(3\bar{n})} < 1$. Then

$$E^{\theta}_{i\delta}(1 - \hat{\mu}_n(\theta)) \leq ((|\textcircled{N}| - 1)\lambda^{-1} + \hat{\mu}_o(\theta)^{-1} - 1)\alpha^n.$$

Proof. Set $\hat{\mu}_o(\theta)^{-1} - 1 = c$, $(|\textcircled{N}| - 1)\lambda^{-1} = c'$. From the representation

$$\hat{\mu}_n(\theta|h) = \hat{\mu}_o(\theta) q_n^{\theta}(h) | \{\textstyle\sum_{\theta'} \mu(\theta') q_n^{\theta'}(h)\}$$

we conclude that $q_n^{\theta'}(h) \leq \alpha^n q_n^{\theta}(h)$, $\theta' \neq \theta$, implies $1 - \hat{\mu}_n(\theta|h) \leq c \cdot \alpha^n$. Hence, we obtain

$$P^{\theta}_{i\delta}(1 - \hat{u}_n(\theta) > c - \alpha^n) \leq P^{\theta}_{i\delta}(U_{\theta' \neq \theta}[q_n^{\theta'} \sigma\eta_n > \alpha^n q_n^{\theta} \sigma\eta_n])$$

$$\leq c'(\lambda^{1/\bar{n}}/\alpha^{1/2})^n = c' \cdot \alpha^n$$

by 4.11 and the definition of α. Now it is easily verified that

$$E^{\theta}_{i\delta}(1 - \hat{\mu}_n(\theta)) \leq c \cdot \alpha^n + c' \cdot \alpha^n.$$

□

If one assumes without loss of generality that $\hat{\mu}_o(\theta) > 0$ for all $\theta \in \textcircled{N}$, then $\sup_{\theta}(\hat{\mu}_o(\theta)^{-1} - 1) < \infty$. The latter bound is minimized by the uniform distribution and then equal to $|\textcircled{N}| - 1$.

REFERENCES

1. Abdel-Fattah, Y.M. (1980): Recursive estimation and control in Markov chains. Faculty of Sciences, University Mohamed V, Rabat, Marocco.

2. Bauer, H. (1978): Wahrscheinlichkeitstheorie und Grundzüge der Maßtheorie. De Gruyter, Berlin.

3. Berge, C. (1963): Topological Spaces. Oliver and Boyd, Edinburgh and London.

4. Bertsekas, Shreve (1978): Stochastic Optimal Control, The Discrete Time Care. Academic Press, New York.

5. Borkar, V./Varaiya, P. (1979): Adaptive control of Markov chains. IEEE Trans. Aut. Cont., Ac-24, 6, 953 - 957.

6. Borkar, V./Varaiya,P. (1980): Identification and Adaptive Control of Markov Chains. Department of Electrical Engineering and Computer Sciences and the Electronics Research Laboratory, University of California, Berkely, California 94720.

7. Derman, C. (1965): Markovian sequential control processes - denumerable state space. J. Math. Anal. Appl. 10, 295 - 302.

8. Doshi, B./Shreve, S.E. (1980): Strong Consistency of a Modified Maximum Likelihood Estimator for Controlled Markov Chains. J. Appl. Prob. 17, 726 - 734.

9. van Hee, K.M. (1978): Bayesian control of Markov chains. Mathematical Centre Tract 95, Amsterdam.

10. Hinderer, K. (1970): Foundations of Non-stationary Dynamic Programming with Discrete Time Parameter. Springer-Verlag, Berlin.

11. Kolonko, M. (1980): Dynamische Optimierung unter Unsicherheit in einem Semi-Markov-Modell mit abzählbarem Zustandsraum. Dissertation, University of Bonn.

12. Kolonko, M. (1981a): The Average-Optimal Adaptive Control of a Markov Renewal Model in Presence of an Unknown Parameter. Preprint No. 430, SFB 72, University of Bonn.

13. Kolonko, M. (1981b): Strongly consistent estimation in a controlled Markov renewal model. Preprint no: 447, SFB 72, University of Bonn.

14. Kolonko, M. (1982): Bounds for the Regret Loss in Dynamic Programming under Adaptive Control. Techn. Report. University of Karlsruhe.

15. Kolonko, M./Schäl, M. (1980): Optimal Control of Semi-Markov Chains under Uncertainty with Applications to Queuing Models. Proceed. In Operations Research 9, 430 - 435, Physica - Verlag, Würzburg - Wien.

16. Kurano, M. (1972): Discrete-time Markovian Decision Processes with an Unknown Parameter - Average Return Criterion. J.Op.Res.Soc.Jap. 15, 67 - 76.

17. Mandl, P. (1974): Estimation and Control in Markov Chains. Adv. Appl. Prob. 6, 40 - 60.

18. Rényi, A. (1964):On the amount of information concerning an unknown parameter in a sequence of observations. Publ.Math.Inst.Hung.Acad. Sci., Series A, 9, 617 - 625.

19. Rieder, U. (1975): Bayesian dynamic programming. Adv. Appl.Prob. 7, 330 - 348.

20. Schäl, M. (1975): On dynamic programming: compactness of the space of policies. Stochastic Processes Appl. 3, 345 - 364.

21. Schäl, M. (1981): Estimation and Control in Discounted Stochastic Dynamic Programming. Preprint no. 428, SFB 72, University of Bonn.

22. Schäl, M. (1982): Asymptotic results for sequential Markov decision models under certainty. To appear.